Managing Health at Work

Other titles from Thomson Professional

Facilities Management
K. Alexander

Clay's Handbook of Environmental Health
17th edition
W.H. Bassett

Environmental Health Procedures
4th edition
W.H. Bassett

Proctor & Hughes' Chemical Hazards of the Workplace
J. Hathaway, H. Proctor and P. Hughes

On the Practice of Safety
2nd edition
F.A. Manuele

Legal Competence in Environmental Health
T. Moran

Occupational Health and Safety Law in the UK and Ireland on CD-ROM
A. Neal and F. Wright

Occupational Health and Safety Law in the European Union on CD-ROM
A. Neal and F. Wright

Environmental and Workplace Safety
J.T. O'Reilly, P.E. Hagan and P. De la Cruz

Statistical Methods in Environmental Health
J.C.G. Pearson and A. Turton

Modern Industrial Hygiene
J.L. Perkins

Safety by Objectives
D. Petersen

Sick Building Syndrome
J. Rostron

Managing Health at Work

A guide for managers and workplace health specialists

Carol Wilkinson
Department of Health and Continuing Professional Studies,
De Montfort University,
Leicester

CRC Press
Taylor & Francis Group
Boca Raton London New York

CRC Press is an imprint of the
Taylor & Francis Group, an **informa** business
A CHAPMAN & HALL BOOK

First edition 1997

Published 2020 by CRC Press
Taylor & Francis Group
6000 Broken Sound Parkway NW, Suite 300
Boca Raton, FL 3487-2742

CRC Press is an imprint of Taylor & Francis Group, an Informa business

ISBN-13: 978-0-419-22980-3 (hbk)

Visit the Taylor & Francis Web site at
http://www.taylorandfrancis.com

and the CRC Press Web site at
http://www.crcpress.com

Typeset in 10/12 pt Palatino by Cambrian Typesetters, Frimley, Surrey

A catalogue record for this book is available from the British Library

Contents

Preface

This book is intended to bring ideas, research developments and suggestions into one volume for managers and workplace health specialists to draw upon for use in their organizations. Some issues raised will be pertinent to specific occupational groups while others are more general. In all, it is designed to be a text that provides stimulus for activity while at the same time highlighting difficulties that may be faced in dealing with specific issues surrounding employee health. One important feature of this book is the emphasis on examples of good practice within organizations around the world.

There are suggestions for further reading and contact addresses for those who wish to pursue a particular subject should the occasion arise.

The first chapter identifies some of the history and background to workplace health developments. It also provides a focus for the need to improve worker health.

Chapter 2 attempts to point out the multi-faceted approaches to the definition of stress and its problems for employees. It provides a series of suggestions for managers, not only in preventing the adverse effects of stress, but also attempts to develop an understanding of the issues surrounding stress.

The subject of back injuries and back pain is discussed in the third chapter. Back pain costs the UK over £3 billion in lost output each year and is thus a major problem for managers in terms of sickness and absenteeism. Prevention strategies are discussed as well as case studies in managing the situations that can arise as a result of employee injury.

Chapter 4 looks at the complex issues surrounding alcohol and the development of drink problems at work. Specific attention is given to the legal position and developments in other countries especially the USA. The chapter ends with a discussion relating to the problem of passive smoking with suggestions for positive action.

Cancer is the second major cause of death next to heart disease. Chapter 5 focuses on occupational exposure to various potential carcinogenic substances such as asbestos and polyvinyl chloride. Information relating to recent developments to reduce the risk of exposure is offered here.

Chapter 6 identifies the risks to employees who may contract HIV and AIDS as a direct result of their work. Particular attention is given to healthcare workers and their attitudes surrounding the condition, and their perception of risk.

Healthy eating and physical activity, although sometimes regarded as a soft option in health promotion, are complex issues when personal circumstances, costs and values are taken into consideration. Chapter 7 outlines recent developments including the process and benefits of dietary counselling, and monitoring and evaluation of health promotion activities at work. Conrad's study in a US corporation is recited here to illustrate the likely reasons for the uptake of physical activity at work.

Chapter 8 touches on the difficult-to-define subject of Repetitive Strain Injury. The chapter looks at recent developments in the definition, specific workplace findings and ways of improving work practices in the quest for injury prevention.

Finally, Chapter 9 identifies the developing role of the workplace health co-ordinator, and provides guidance to managing health at work.

Carol Wilkinson
May 1996

Acknowledgements

The inspiration for this book was developed during my working years in South Wales. I am indebted to those people who gave me an entrée into their organizations and assisted me in developing the workplace co-ordinator role. Special thanks to John Newcombe, Lliw Valley Borough Council, The South Wales Constabulary Health Promotion Group, and staff of the West Glamorgan Probation Service.

I am grateful to Rhiannon Urqhart, Health Promotion Wales for her support and guidance, and to Jacquie, Sandy, Cindy, Ange, Elaine, Ben and others of The Unit, for their warmth and good humour.

Focus on workplace health 1

Background to workplace health developments • Why the USA? • What about the United Kingdom? • What is workplace health promotion? • Rationale for workplace health improvement • Barriers to workplace health improvement • Make Health Your Business Awards, Wales, UK • Managing Health Improvement Project • Legislation • Suggested further reading • References • Useful addresses

BACKGROUND TO WORKPLACE HEALTH DEVELOPMENTS

Concern for worker health improvement in modern times together with involvement in reform started in England and Scandinavia during the 1960s. This has occurred for the most part, as a result of union activity and subsequent developments in legislation. Indeed, Scandinavian researchers have developed strategies to improve worker health which became enshrined in legislation during the late 1970s. Sweden today leads the way in research related to stress in the workplace and its impact on its employees. The contributions are largely due to the endeavour of the Scandinavian School of workplace health researchers, Gardell, Karasek and Theorell, to name but a few. However, much of the impetus for improvement came from United States corporations and businesses during the 1970s when 'Wellness' programmes were introduced. These programmes were devised to assist employees who were suffering from strain and the resultant misuse of alcohol and drugs.

Companies participating in these ventures included Lockheed, Johnson & Johnson, IBM, Kimberly Clark, AT & T and Ford Motor Company in the initial stages. By the 1980s up to 40% of larger companies had introduced workplace health programmes.

Looking after employee health has long been a concern of personnel managers and safety representatives, the main provision being first aid, insurance packages, sickness benefit and general support. Some large organizations have engaged the services of occupational health practitioners, giving medical treatment, care and guidance to the employee experiencing injury or sickness.

The Japanese culture and paternalistic management style have led to extensive provision of workplace health promotion initiatives. The concept of wellness has been consistent with Japanese company values of quality control where problems are identified at an early stage and dealt with rather than merely detected (Opel and Clark, 1988; Schonberger, 1982).

There are marked differences in work culture between Japanese and UK companies, especially in relation to the problem of absenteeism. Japan's rate of absenteeism is much lower. Indeed, the Industrial Society in 1993 reported the national absenteeism rate in the UK (for 1989) was 3.97% compared to 2.35% for Japanese companies operating in the UK. The highest rates were found in the public sector, with the NHS fairing worst at 5.94%. Common problems identified were stress, sickness in the family, workplace accidents, and alcohol and drug problems.

Studies in a number of companies have demonstrated that the reduction of absenteeism is possible.

The Canada Life Study found that participation in programmes relating to worker health improvement, led to a 22% reduction in absenteeism and a 90% reduction of staff turnover (Cox, 1978). Bertera (1990) also studied absenteeism among full-time employees in a large, multi-location company in the USA. Blue-collar employees participating in a comprehensive health promotion initiative showed a 14% decline in absenteeism over a two-year period. When measured against costs, reduced absenteeism offset total programme costs within the first year of intervention and provided a return of $2.05 for every dollar invested in the initiative by the end of the second year.

The concept of workplace health improvement for employees, with the exception of Scandinavia is in its embryonic stages. Initiatives taken forward by senior management have focused on beneficial change in personal health behaviour and subsequent reduction in associated medical care costs. However, the concept of workplace health improvement for employees is relatively new in Europe, with much concentration on cardiovascular disease programmes.

This is changing.

The European Foundation for the Improvement of Living and Working Conditions, founded in the late 1980s, has identified the aim of worker health improvement as one of its necessary tasks. The formulation of a corporate culture through policy development is another. To date, the Foundation has conducted research and development programmes within the European Community which have highlighted the patchiness of initiatives, some of which are indicated below.

Greece
The programmes offered (in the workplace) are largely traditional in their approach. For example, the identification and assessment of health hazards, medical examinations, first aid and emergency services. Smaller companies do not have a high level of activity in the health and safety area. Also, there is concern that the agricultural sector, employing up to 30% of the workforce, does not have any systematic health and safety policy. The unregulated use of a range of fertilizers, pesticides and weedkillers is cause for considerable concern.

Italy
Health service staff devoted to general health promotion constitute just 0.32% of the entire personnel of the health service. There are scarce resources. Preventative action is limited. To date, five out of the seventeen regions in Italy have now set up boards to look into prevention strategies in health and safety at work.

(EFILWC, 1989)

In the United Kingdom, the Managing Health Improvement Project (MAHIP) has been developed and is now (1995–96) being piloted in a number of health authorities and trusts. The Confederation of British Industries (CBI), Trades Union Congress (TUC) and Health and Safety Executive have also joined the Department of Health in spearheading a Health of the Nation Strategy to improve workplace health in medium-sized and large organizations. There also exist incentives in Wales to improve worker health through the CBI's 'Make Health Your Business Awards'.

Perhaps most significantly, Health At Work was introduced in the NHS during the summer of 1992 as part of the UK's Health of the Nation Strategy. This initiative was launched with two aims:

1. To introduce a systematic approach to promoting health throughout the National Health Service.
2. To engage all staff in health promoting initiatives.

The project was designed to build on the 'Look After Your Heart' (LAYH) initiative and to develop a long-term strategy to benefit the whole workforce. The initiative addresses environmental issues, looks at working practices, the development of support networks and focuses on lifestyle. Much of the development for this project to date has been in the south-west and western regions of Britain.

Control of adverse stress at work has been identified as critical to the physical and psychosocial well-being of the employee. Noise, heat, exposure to chemicals, the adverse effects of shiftwork, highly repetitive tasks, poor or overbearing management and supervision, technological change without training and increasing workload have all been seen as factors that impinge upon employee well-being.

The workplace health movement in the USA has continued to lead the way.

WHY THE USA?

The drive towards improving employee health in the USA came from three different standpoints:

1. cost containment;
2. lifestyle risk factors;
3. cultural wellness.

COST CONTAINMENT

This was regarded as the key to corporate health policy. Managers envisaged that if companies could reduce their medical insurance and disability claims, then this would lead to lower healthcare costs and paying of sickness benefits, reducing operational costs. Also, employee productivity is seen as crucial within a competitive market. Keeping workers healthy means greater productivity.

LIFESTYLE RISK FACTOR

The medical profession emphasises that there is scientific justification for implementing strategies to improve employee health. Medicine sees individual lifestyles and behaviours in terms of risk factors which are central to the development of chronic disease. Risk factors are perceived by the medical profession as behavioural and so can be modified, reducing risks to health. In 1964, the US Surgeon General's Report on Smoking implicated smoking at work and more generally, as being associated with the development of cancer. The Framingham Heart Study has linked high cholesterol, smoking and hypertension to the risk of heart disease. More positively, the researchers indicate that certain behaviours and habits can make substantial contributions to health status and longevity.

This approach has been criticized on the grounds that it overstates individual responsibility, is victim blaming and often provides a self-serving justification for medical services and medical research. Nonetheless, it is valuable in indicating that employee health can improve through practical measures implemented in the workplace.

CULTURAL WELLNESS

Health interests in the US have been promoted by cultural and social values rather than scientific advances. Workplaces encouraging jogging in the late 1980s saw over 60 million participating between the ages of 20 and 50 years. This for the most part has been among the professional middle classes who have also stimulated the sales of health foods, participated in exercise classes at work, and accounted for increased attendance at health and fitness clubs.

In conclusion, progressive employers, and demand from certain sections of the population and the medical profession have driven programmes for health improvement among employees in US corporations.

WHAT ABOUT THE UNITED KINGDOM?

In the UK, the current workforce consists of approximately 22 million, three million of whom are self-employed. There are at least 12 000 businesses (0.5%) which employ 200 or more people; 96% of all workers (45% of the workforce) are employed in small businesses consisting of less than 20 people. This figure is set to increase with the rapid expansion in small businesses (although at least a third fail within three years of commencement).

Many of them are well versed in the legislative requirements for the maintenance of health and safety, but few take note of the underlying causes to worker dissatisfaction and ill health.

A recent survey carried out by the CBI (Department of Health, 1995), estimated the direct costs of sickness absence to employers annually is in the region of £11 billion. The major causes of sickness absence are:

- **alcohol and drink-related illness** costing £1.7 billion, a total of 8 million working days lost per year;
- **mental illness** costing business £3.7 billion per year, a total of 91 million working days lost per year;
- **coronary heart disease and strokes** costing £2.1 billion, a total of 62 million working days lost per year.

These figures indicate an urgent need for action to improve employee health. Progressive employers have undertaken excellent programmes in this field; indeed this book outlines some of the more notable examples from around the world to assist managers in improving their work practices and encouraging employees. The UK has its own projects, some of which are described in the chapters of this volume, while others still need to be brought into the public domain. It is hoped that this book will encourage health practitioners and progressive managers to develop innovative practice.

In 1992 The Workplace Health Taskforce was set up in conjunction with representatives of industry and health. They include the Confederation of British Industry, the Health and Safety Executive and the Wellness Forum. The aims of the Taskforce are to look at current provision, to market health promotion at work, to evaluate the effectiveness of initiatives in workplace health, as well as to identify employers' responses to the Health of the Nation Targets.

Findings and recommendations of the Taskforce are used to:

- ensure greater commitment from central government;
- set up a steering group to oversee work;
- establish local networks to market health promotion;
- disseminate information related to workplace health;
- encourage employers to undertake needs assessment, and to define objectives and ways of monitoring and evaluating service provision against health benefits.

Its main concentration to date has been the marketing and evaluation of programmes carried out by companies taking up the initiative.

WHAT IS WORKPLACE HEALTH PROMOTION?

Parkinson (1982) identifies workplace health promotion as

> a combination of educational, organizational and environmental activities designed to support behaviour conducive to the health of employees and their families.

Traditional programmes have consisted of health education and screening and/or intervention designed to change employees' behaviour in a healthy direction and to reduce associated risks. Activity has ranged from single interventions such as screening for hypertension to more comprehensive wellness programmes. These have included risk assessments, aerobic exercise, fitness classes, smoking cessation, nutrition and weight control, stress management, healthy back programmes, cancer prevention and screening, drug and alcohol abuse reduction and accident prevention.

As well as concentrating on individual employee behaviour, it is crucial that managers take a lead both in terms of getting the message across by example, and by providing funding and resources so there is a coherent organizational approach. Employees will know that there is a serious intent to drive any initiative forward if there is managerial support from the outset. It is not just employee behaviour that needs attention but also the culture of the organization. Managers are in the ideal position as change agents to affect this.

RATIONALE FOR WORKPLACE HEALTH IMPROVEMENT

Workers spend over 30% of their waking hours at the workplace. They are a captive audience for any health promotion intervention. Rosen (1984) notes that corporate executives and managers are attracted by evidence of:

- improving employee health and fitness;
- reducing absenteeism;
- decreasing medical and disability costs;
- decreasing staff turnover;
- improving employee mental alertness;
- improving morale and job satisfaction;
- enhanced corporate image;
- improving productivity.

This has been demonstrated in companies such as Dupont and Johnson & Johnson for which evidence of substantial gains is available.

TYPES OF INTERVENTION

Workplace health improvement has largely developed in three distinct areas:

1. **motivation and incentive strategies;**
2. **education and training;**
3. **environmental and social support.**

MOTIVATION AND INCENTIVE STRATEGIES

Many workplaces have introduced a variety of initiatives via their occupational health departments or independent organizations including:

- health screening, e.g. breast cancer, cervical cancer, and cholesterol check;
- health risk appraisal;
- incentives for behaviour change and participation, e.g. health fairs, contests and lotteries.

EDUCATION AND TRAINING

The range of activities has been wide and has included:

- the use of self-help materials;
- computer-based instruction;

- individual counselling;
- classroom instruction and group discussion;
- company presentations.

ENVIRONMENTAL AND SOCIAL SUPPORT

Examples include:

- cafeteria menu changes;
- no-smoking policies;
- involvement of family members;
- provision of accident prevention materials;
- infection control policies.

BARRIERS TO WORKPLACE HEALTH IMPROVEMENT

Both manager and employee are likely to have reservations relating to health improvement projects at work for different reasons. An individual is primarily employed to assist in meeting the goals of the organization and to perform duties efficiently within the parameters set. Managers ensure that certain standards are met in the performance of employee tasks and that, for the most part, the company makes a profit.

For managers the main reservations may include:

- resistance to change;
- indifference to health issues;
- concern about resource implications;
- competing priorities, e.g. disruption in productivity.

For employees reservations may include:

- interference in personal life;
- confidentiality;
- diversion of attention from competing issues;
- release time and logistics of programme delivery.

MAKE HEALTH YOUR BUSINESS AWARDS, WALES, UK

The Make Health Your Business award was an initiative developed in the early 1990s by Health Promotion Wales, and endorsed by the TUC and CBI Wales. Its aim has been to encourage employers to develop initiatives for improving employee health at work. The awards are given on a yearly basis to the company which has made the greatest effort to improve facilities and promote employee health. Companies have to provide written documentation and practical evidence to demonstrate how they have or are seeking to address issues relating to:

- healthy eating;
- environmental care;
- safety;
- cancer prevention;
- sensible drinking;
- healthy heart;
- smoking cessation;
- physical activity;
- healthy sexuality.

Successful companies have included DVLA (Swansea), Powys County Council, Kelloggs, Welsh Water, the University of Glamorgan and Midland Bank.

The initiative is currently undergoing an alteration in its status and philosophy and will be re-launched during the latter part of 1996, underlining its **Corporate Health Standard.**

MANAGING HEALTH IMPROVEMENT PROJECT

This project was undertaken during the early part of the 1990s at the Department of Education, University of Keele. It was jointly commissioned by the Health Education Authority and the NHS Training Directorate. Its aim is to generate training materials in health promotion and health improvement for management employees in the National Health Service. (The project is directed by Professor Richard Kempa and assisted by Dr Ina Simnett.)

The project is linked into the government's Health of the Nation Strategy. Materials have been developed for use at a local level which identify with the targets set out in the strategy.

The main themes include:

- health policy, needs assessment, priorities and targets for health gain;
- management competencies for health improvement;
- quality assessment;
- design and implementation of strategies for health gain.

LEGISLATION

Throughout the following chapters, the legal requirements of managers are outlined relating to specific issues concerning for example, alcohol consumption in the workplace, employee protection against hazardous substances, and assisting the employee with HIV to maintain an appropriate level of work activity.

There is a list at the end of Chapter 9 indicating the legislation relevant to managers in the UK.

SUGGESTED FURTHER READING

Alexander, J. (1988) The ideological construction of risk: an analysis of corporate health promotion programs in the 1980s, *Social Science and Medicine*, **26**(5), Pergamon Journals Ltd, 559–67.

Bertera, R.I. (1990) The effects of workplace health promotion on absenteeism and employment costs in a large industrial population, *American Journal of Public Health*, **80**(9), 1101–5.

Conrad, P. (1988) Health and fitness at work: a participant's perspective, *Social Science and Medicine*, **26**(5), Pergamon Journals Ltd, 545–50.

Department of Health (1995) *The Health of the Nation: ABC of Health Promotion in the Workplace, A Resource Pack for Employers*, Department of Health.

Ewles, L. and Simnett, I. (1995) *Promoting Health: A Practical Guide to Health Education*, 3rd edition, John Wiley.

Paoli, P. and Litske, H. (1994) *Monitoring the Work Environment*, Report of the Second European Conference, European Foundation for the Improvement of Living and Working Conditions.

REFERENCES

Bertera, R.I. (1990) Planning and implementing health promotion in the workplace: a case study of the Du Pont company experience, *Health Education Quarterly*, **17**(3), 307–27.

Cox, T. (1978) *Stress*, London: Macmillan.

Department of Health (1995) *The Health of the Nation: ABC of Health Promotion in the Workplace*, A Resource Pack for Employers.

EFILWC (1989) *Workplace Action for Action Health: A Selective Review and a Framework for Analysis*, Working Paper No. EF/WP/89/30/EN.

Opel, G.L. and Clark, M.W. (1988) Sport, fitness and recreation in the Japanese workplace. In E.F. Brown, R.A. Clumper, B.B. Pendleton and C.A. Pooley, *Comparative Physical Education and Sport*, Vol. 5, Human Kinetic Books, 285–92.

Parkinson, R.S. (1982) *Managing Health Promotion in the Workplace*, Mayfield, Palo Alto, California.

Rosen, R. (1984) Worksite health promotion: fact or fantasy?, *Corporation Commentary*, **1**(1), 1–8.

Schonberger, R. (1982) Production workers bear major quality responsibility in Japanese industry, *Industrial Engineering*, **14**(12), 34–40.

USEFUL ADDRESSES

Health Education Authority, Hamilton House, Mabledon Place, London WC1H 9TX.

Health Promotion Wales, Ffynon-las, Ty Glas Avenue, Llanishen, Cardiff CF4 5DZ.

Confederation of British Industries (CBI), Centre Point, 103 New Oxford St., London WC1A 1DU.

Stress 2

What is stress? • Defining stress • How does the body react under stress? • Problems of stress at work • Causes of stress • The work environment • The job • The organization • Other causes of stress at work • Recognition and effects of stress at work • Managing organizational stress • Organization of the work environment • Organization of job tasks • Promoting health in the workplace • Management action plan • Guidance for managing stress at work • Individual coping strategies • Work management • Coping with organizational stress • Summary • Suggested further reading • References • Useful addresses

WHAT IS STRESS?

- Stress covers a wide range of views, thoughts, feelings, body conditions, health, behaviour and work-related issues.
- Stress is a complex phenomenon.
- Stress is a useful concept for the understanding of human functioning and performance.
- Stress is usually regarded as unpleasant, undesirable and unwanted but not always; some stress can have positive effects.
- Stress underlies many common concerns, e.g. anxiety, anger, depression, colds, headaches, boredom, absenteeism.
- Stress is highly individual in its expression.
- Stress occurs when significant internal or external demands are experienced as exceeding available human resources.

DEFINING STRESS

Studying stress is for the most part a process which involves observation of behaviour. It is by no means an easy task; as with most processes there are a variety of actions and responses which are not always predictable. However, there are certain characteristics which continue to appear in definitions forwarded by experts in the field:

- Stress is a consequence of interaction.
- Stress is an adaptive behavioural process.
- Stress is a natural part of human functioning.

Seyle (1978) who did much pioneering work on the subject of stress, defined it as a 'non-specific (adaptive) response of the body to any demand made upon it'.

Whether the problem is provoked by a quarrel or trying to cope with overwork the effect is the same. Individuals feel stressed and some adaptation is required to adjust to it. He called this defensive adjustment to stress the **General Adaptation Syndrome (GAS)**:

- **general** in that many parts of the body are affected by 'stressors', i.e. that which causes a stressed response, for example, receipt of bad news, excessive noise, a smoke-filled room, moving house, threat of redundancy;
- **adaptation** because the stressor stimulates bodily defences in an attempt to cope with the stress;
- **syndrome** because several symptoms or reactions are produced simultaneously.

There are three stages of GAS: alarm, resistance and exhaustion. In the **alarm** state, the stress evokes 'fight and flight' responses as the body mobilizes to meet the threat imposed by the stressor. The brain responds by sending biochemical messages to the respiratory, circulatory and muscular systems causing breathing, heart rate and blood pressure to increase, muscles to tense and contract and pupils to dilate. Additionally, blood sugar, adrenalin and cholesterol levels increase.

The **resistance** phase develops following exposure to stress for any length of time. The individual begins to adapt to the stressor stimulus, but the ability to cope with further stimuli decreases. Resistance may rise above the tolerable norm. At this point the individual begins to be generally overwhelmed by feelings of fatigue, anxiety and tension.

Prolonged exposure to stress means the body has become somewhat adjusted to the stressed state. Consequently, all the available energy is used up and the individual reaches **exhaustion.**

Individuals only have finite reserves of energy with which to fight stress. Once their resources are depleted, they may become seriously ill, especially if any additional burdens which add to the stressed response is experienced. GAS places excessive wear and tear on the body each time it is activated. The result is general lassitude, ageing and illness, especially if it reaches the stage of exhaustion.

HOW DOES THE BODY REACT UNDER STRESS?

The body under stress, enters the 'fight and flight' response. The nervous system secretes the hormones adrenalin and noradrenaline to prepare the other systems of the body to react efficiently. The hormone acts as a powerful stimulant to:

- increase the heart rate;
- increase blood pressure;
- raise blood sugar levels;
- raise metabolism;
- speed up reflexes.

The result is increased short-term capacity and performance as more blood is carried to the muscles and lungs. Energy supplies are boosted and responses sharpened. When this energy release is not translated into instant action, the long-term effect can lead to cardiovascular disorders such as heart disease, strokes, kidney problems as a result of high blood pressure, and even the aggravation of diabetes. Some diabetics may even experience irregular hypoglycaemic difficulties.

Another hormone, thyroxine, is also released under stress. Its function is to boost the body's metabolism and increase the rate at which energy can be burnt and translated into physical energy. A prolonged racing metabolism leads to exhaustion, weight loss and eventual physical collapse.

Cholesterol is also released into the bloodstream. This increases energy and aids muscular function. A permanently elevated blood cholesterol greatly increases the risk of arteriosclerosis (hardening of the arteries), a major cause of heart disease. Blood vessels constrict and blood thickens, so that it flows less rapidly and coagulates more quickly in the event of flesh wounds. The heart has to work harder to pump blood through narrowed vessels, increasing the likelihood of heart problems such as chest pain, breathlessness and palpitations.

The body's natural analgesia known as **endorphins** are released into the blood to reduce sensitivity to bodily injuries. When endorphins are depleted, the body becomes more sensitive to aches and pains, hence the experience of headaches, backache and increased bodily tension under stress. The hormone **cortisone** secreted from the adrenal glands is responsible for shutting down allergic reactions which can interfere with breathing under stress. Once the immediate effects wear off, allergic reactions return with increased force; this can increase the likelihood of asthma attacks.

In addition to the secretion of hormones and chemicals during the 'fight/flight' response, other reactions occur within the body, such as the following:

- A shut down of the digestive system allows blood to be diverted from the stomach so it can be used by the lungs and muscles. Secretion is also arrested so the mouth becomes dry because of lack of saliva. Prolonged shut-down of digestion can lead to digestive difficulties and increases the likelihood of peptic ulcers.
- Blood is diverted from the surface of the skin for use elsewhere, making it pale. Sweat production increases so cooling the muscles. Excessive sweating can lead to body odour and excessive heat loss.
- The air passages dilate allowing greater uptake of oxygen and an increased breathing rate. Super-oxygenated blood can lead to dizzy spells, blackouts and upset heart rhythms.
- The senses become more acute under stress and mental performance improves, producing enhanced short-term functioning. If stress is prolonged, these effects go into reverse, actively depressing sensory and mental responses.
- Finally, sexual hormone production is reduced. This avoids the diversion of energy or attention to sexual arousal. When prolonged, the reduction can lead to impotency, frigidity, and other sexual difficulties.

So stress causes physical and behavioural change. What are the consequences in the work setting?

PROBLEMS OF STRESS AT WORK

In his study of stress and behaviour in organizations, McGrath (1976) identified a relationship between stress and job performance in terms of the underload/overload continuum. (See Figure 2.1.)
There is a point along the continuum where people are at their best in terms of performance. They possess high energy, have sharp perception and exude calmness. If they reach a stage of imbalance, that is, do not have enough work to do or have too much work, then the stress becomes problematic.

Beehr and Newman (1978) identify work stress as a situation wherein job-related factors interact with a worker to change, disrupt or enhance, his or her psychological and physiological condition. As a consequence, the person's mind and/or body is forced to deviate from normal functioning. This very much equates with McGrath's findings in that to under-employ or to overload the individual is to prevent full concentration on the task in hand and reduce performance. This is a frequent consequence of employees who are not informed clearly of their role if starting a new job with the company, or taking on additional tasks for a sick colleague with little or no support/relief.

Stress and job performance of the individual

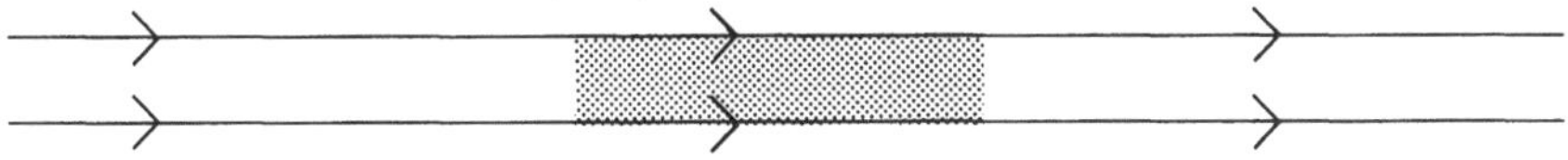

Low	Optimum	Low
UNDERLOAD		**OVERLOAD**
• boredom	• high performance	• insomnia
• decreased motivation	• high energy	• irritation
• absenteeism	• sharp perception	• errors
• apathy	• calmness	• indecision

(Source: Adaptated from McGrath, 1976.)

Figure 2.1 Underload/overload continuum

Stress in all its forms is regarded as a major threat to personal well-being in the advanced industrial world. In addition, the problems associated with stress at work are connected to the effects on:

1. the business and
2. the individual.

THE BUSINESS

- Stress is a major contributor to absenteeism and sickness absence. Conservative estimates have put the total cost of stress to industry at over £26 billion a year in terms of lost hours, accidents and sub-standard work. As much as 2% of the UK's GNP is lost through stress-related problems. This figure takes no account of the cost of absence that has been certified under another condition as few people want a certificate referring to their psychological status or of absence that lasts less than seven days which is subject to self-certification.
- Of the 360 million days lost annually through sickness in the UK, an estimated 50% are due to stress-related absence. Thirty days are lost to stress for every single day lost to industrial disputes.
- Gardell and colleagues (Gardell, 1977; Gardell *et al.*, 1982) have conducted a variety of studies in the workplace that demonstrate that the main factors perceived to be related to absenteeism were poor motivation and monotony.
- In California, at least 50% of claims by employees during the 1980s were for stress-related disorders and an estimated quarter of all medical insurance claims by employees are for emotional illness.

THE INDIVIDUAL

- In 1990, the European Foundation carried out a survey indicating that of the 150 million workers in the European Community, 23% work to tight deadlines all the time; 23% also work on short repetitive tasks and 38% have no possibility of changing tasks or work methods; 35% are not able to change the speed or rate of work.
- The cost of stress to individuals has many consequences. These include absenteeism, anxiety, alcohol abuse, depression, anger, gambling, theft, sexual difficulties, family conflicts, heart disease, poor concentration, impaired judgement and poor decision-making.
- General practitioners in the United States estimate that 70–90% of the health problems presented each year are related to excess stress. In any given year, it is estimated that 25% of the US population experience severe emotional stress, and it is an active problem for at least 5% to 10% of the workforce.
- The UK has one of the highest rates of heart disease in the world. It kills 25 000 men and 7000 women before their 65th birthday. Managers have a 1 in 4 chance of being struck by stress-related heart disease before they reach retirement.
- A group of working women were monitored over a ten-year period for signs which indicated the development of coronary heart disease (CHD). The results showed that the combined effect of highly demanding work and a lack of clarity in direction from their supervisors was a powerful predictor of CHD in women in clerical jobs. There was a clear exposure–response relationship. CHD rates rose from 2.4% in the low demand/high clarity group to 31.3% in the high demand/low clarity group. These associations persisted when controlled for the four standard coronary risk factors of age, blood pressure, blood cholesterol and cigarette smoking (Webb *et al.*, 1989).

CAUSES OF STRESS

The work environment has been identified as one of the main sources of stress. Its most marked affects occur if:

- employees have no degree of influence and control over their work;
- employees perceive the work to be meaningless and not worthwhile;
- employees fail to develop an affinity with fellow workers as a social support system and fail to identify with it through their work.

As people spend a great deal of time in the workplace, we need to look much more closely at why excessive stress is experienced there. We also need to look at other factors that impinge upon working life. For ease of explanation, factors associated with the causes of stress may be grouped as follows:

- **The work environment.**
- **The job.**
- **The organization. culture and management styles.**
- **Factors intrinsic to work.**

THE WORK ENVIRONMENT

HAZARDS

The following is a series of factors which at times may be insufficiently or totally unregulated and which can lead to difficulties ranging from moderate discomfort to ill health.

- The building's temperature is inadequate for normal functioning, i.e. too hot or too cold or fluctuating temperatures.
- Employees may lack protection from excessive noise (although protection is required under the Health and Safety at Work Act).
- There may be vibration present that interferes with regular activity.
- Poor or intense lighting can lead to mistakes being made in performing work tasks exposing individuals to accidents.
- Ventilation may be inadequate.
- There is exposure to toxic substances, chemicals, dust, gases and fumes.

SMOKING

A smoke-filled atmosphere can add to discomfort. Workplaces with unregulated smoking and with no agreed smoking policy is generally unsatisfactory to all. Smoking is both a cause and a result of stress. People under stress smoke more. Non-smokers in a smoky atmosphere can develop bad personal relationships and are distracted from their work.

BUILDING CONDITION AND LAYOUT

Buildings which are too old or poorly maintained, unsuitably designed, overcrowded and untidy are uncomfortable to work in and may also be a safety hazard. Work areas that are too small or overcrowded may not provide enough room to carry out the work properly and allow no privacy. Some layouts make workers feel cut off from each other and isolated. Although they appear to facilitate comfort and communications, open-plan layouts can also cause stress through being noisy and a source of tension if employees need a quiet environment to complete that all-important report for their superior.

STAFF FACILITIES

Overcrowded restrooms and dilapidated cloakrooms do not provide proper relief for employees. They may become an additional source of stress if staff cannot have a comfortable break or if they have to worry about the safety of their possessions.

MACHINERY AND EQUIPMENT

Studies have shown the following to be stressful.

- Machinery that is awkwardly placed or badly sited.
- Equipment that is not suitable for the job.
- Equipment not designed to suit the user; desks, chairs and machinery can cause stress and strain if they force the user to adopt an awkward, uncomfortable position for long periods.
- Machinery and equipment that is old or in poor condition, that continually breaks down and is not maintained or regularly serviced.
- Overlong exposure or glares from visual displays can contribute to stress.
- Lack of financial resources within the organization can mean equipment and the required facilities are unavailable to carry out the functions required within the job.

THE JOB

> A study of neonatologists indicated that major dissatisfactions in the workplace were over work in caring for a number of sick patients, lack of resources, inadequate salary for the work done, and too many administrative demands.
>
> (Clarke *et al.*, 1984)

JOB DESIGN

Many manufacturing processes today have been reduced to smaller tasks requiring little skill or initiative. This can be rather irksome for individuals who need to carry them out as they are tedious and often repetitive.

SHIFTWORK AND INTERNAL ROTATION

It has been estimated that over three million full-time workers are employed on shifts in the UK. Shiftwork disturbs sleep patterns and causes loss of sleep, disrupts family and social life and biological rhythms.

The body's natural rhythm is known as the **Circadian rhythm**; over the 24-hour period there is a time to sleep, time to work, a time at which the body's metabolism is at its peak and a time at which it operates more efficiently and when thought processes are at their peak. There is also a time at which the mind and body needs to rest and recuperate. In addition to these major peaks and troughs there are a number of minor ones spread over 24 hours. These rhythms are set in part by nature which intends people to work from dawn to dusk and rest during the hours of darkness. Society and work demands that people start and finish their day rather later than the pace of their body's natural rhythm.

Shiftwork and abrupt fluctuations from day to day, as in business, factory, police, hospital and social work to give but a few examples, are circumstances where people have to cope with sudden crises at all hours. This can trigger both psychological and physiological stress.

> A study of night-shift workers in the manufacturing plant of a cement industry in Lower Austria suggested that psychosocial and environmental stress factors at work act independently from shift-related stress factors.
>
> (Cervinka, 1993)

Although the rate of accidents at work is similar between day and night workers, accidents at night are more severe. The debilitating effects that shift work has brought about has led to its ban in a number of industries in Norway, Sweden and Belgium.

WORKLOAD

Surveys of people working in financial institutions in the UK have found that, of the 64% who said their organizations were stressed, the majority said too much work was the major cause. Too much work can be the result of changes in the organization, in particular when staffing levels are reduced and individuals find themselves expected to carry out work which would previously have been done by two people, or when job requirements alter and retraining is inadequate or not provided at all.

Overwork can arise from chronic staff shortages. Managers may have far too high an expectation of their staff, and be insensitive to the impact overwork and long hours have on work and family relationships. Many workers have little or no choice in determining either their hours or the amount of work they will have to do. This is because they have to respond to the needs and demands of others. For example, the majority of office and clerical workers have to deal with workloads created by their bosses. Doctors and nurses responding to emergencies experience unpredictable workloads which they need to deal with before moving on to the next set of tasks. Other workers have to cope with peak periods which can overload them, e.g. railway ticket clerks dealing with the morning and evening rush hours; telephonists answering the peak of enquiries mid-morning; bus drivers in the rush hour. In slack periods they have little to do, which can be as stressful as having too much.

Jobs with outstanding demands make continual physical and mental challenges on the individual, eventually leading to stress reactions. Physical overload can result from operating poorly designed machinery, sitting in badly designed chairs or typing positions or standing for long periods. Mental overload may arise from several sources including the necessity of absorbing a great deal of information in a short space of time, or having frequent contact with members of the public who may themselves be experiencing excess stress. This is experienced particularly in the caring professions and leads to a situation where workers are unable to act effectively and demonstrate an inability to finish a task.

ROLE CONFLICT

This arises when an individual is subjected to conflicting or contradictory job demands. Managers who have to co-ordinate work between different departments or between the company and the outside world can find themselves expected to fulfil incompatible demands and become highly stressed as a result. They have insufficient resources to carry out all aspects of their jobs properly and become forced to make arbitrary decisions about priorities. They may find that rather than co-operating with their colleagues, they have to compete with them for more workers or a larger budget.

Individual workers may find that their personal values come into conflict with the values of their boss or organization. Women at work can be especially vulnerable to the stress that arises from role conflict. Many women perform the dual role of running the home and maintaining a job or career. Many women have to take jobs which are part time, insecure and with few opportunities for promotion.

Poor pay, status and promotion prospects can lead to stress. Professional people regard their identity as related to their work. If the job is poorly regarded by society in general or by their paymasters, it is difficult to stop this rubbing off on the individual's sense of who s/he is and how significant s/he is. Personal worth is lost if other people seem to regard what they are doing as of minor importance or as the sort of thing that anyone with minimal training and ability can do.

ROLE AMBIGUITY

> A survey of 1000 certified dental assistants in Ontario, Canada indicated that approximately one-third had been bothered by stress at work on three or more days in the previous week. The main sources of stress were working under constant time pressures, running behind schedule and feeling undervalued by their superiors. Analysis of the survey results revealed that the main predictors of work stress were not having a clear job description, working long hours, domestic pressures and age.
>
> (Locker *et al.*, 1989)

When an individual is not entirely sure of their role in the organization there can be difficulties. The lack of clarity comes from several sources:

- lack of induction/briefing/training;
- lack of knowledge about the job. This can lead to inefficiency, duplication of roles, and individuals working above or below their level of competence;
- lack of awareness regarding the scope and purpose of the job, how it fits with the jobs of others, how to judge whether they have done a good job and how much support to expect from their boss. This leads to uncertainty and insecurity, taking the individual away from familiar landmarks, sense of self and how to orientate themselves. A major contributory factor here also is poor communication and poor planning.

Such confusion can be the result of many things:

- poor selection;
- insufficiently clear or differentiated job description;
- inadequate communications and co-ordination;
- lack of knowledge and experience on the part of the manager.

THE ORGANIZATION

CULTURE

All companies have shared values and ways of behaving. This may be a source of stress because an organization's value system may come into conflict with personal attitudes and beliefs in various ways. If the company places continual pressure on its employees to succeed at all costs, demands absolute commitment to the job and only rewards those who are 'workaholics,' conflicts will inevitably develop between work and home life. This may become intolerable if family pressures conflict with work pressures.

MANAGEMENT STYLES

A manager's presentation is crucial to the success and ethos of a company. The manager influences every area of the company's work from planning and policy to communications and co-ordination. S/he also sets the tone of interpersonal relationships throughout the organization.

Excessive problems leading to unwarranted stress can develop if a manager:

- treats employees unfairly;
- does not give time to others;
- never motivates others or gives praise for good work;
- never encourages feedback;
- lacks an open-door philosophy;
- lacks direction;
- is pompous and 'hides' behind policy.

Autocratic management can cause major problems for employees. This can encourage interpersonal competition, a rigid hierarchy, impersonal communication, lack of and poor communication, decisions made without consultation, a narrow focus on individual achievements, and employees will have few opportunities for creativity.

Such managers exhibit:

- poor communication or no communication of the organization's goals;
- self-interest at the cost of employee's well-being;
- no accountability to fellow employees;
- decisions and actions taken without consultation.

OTHER CAUSES OF STRESS AT WORK

DISCRIMINATION

Women, people of a different culture, religion, nationality or racial group and disabled people are liable to be discriminated against at work. Working environments that encourage the stereotype of the fit and healthy white heterosexual male as 'normal' and the only truly acceptable model can make life intolerable for an 'outsider'.

For women, a manager using sexist language and entertaining behaviour that may be regarded as harassment leads to excessive stress as it is degrading and offensive.

Stress for women at work arises through:

- home/work conflicts (dual role problem);
- low expectations by employer;
- fear of success;
- tokenism;
- mobility problems;
- childcare;
- male-dominated organization/management;
- stereotyping;
- superwoman syndrome;
- high personal expectations;
- lack of assertion;
- few training opportunities;
- promotion blocks.

Black people often experience personal abuse and antagonism. Institutionalized racism means that despite the existence of equal opportunities policies, many will be restricted to low pay and insecure jobs, and poor promotion prospects. Those fortunate in achieving managerial positions within the organization are often in place as a token gesture and their influence regarded with disdain or having little effect.

The disabled who find it difficult enough to obtain employment, have to cope with frequent well-meant and often misguided attempts to 'help' them which can take away their independence. Prejudice and harassment can cause considerable stress for individuals in these groups as well as undermining constructive personal and professional relationships throughout a department or entire organization.

RESPONSIBILITY AND CHANGE

People with a high degree of responsibility but whose authority is limited are more likely to suffer from the effects of stress.

Some experiences include:

- lack of control or involvement;
- insufficient mental stimulation;
- high noise levels;
- repetitive and dull tasks;
- blocking decisions and frustrating attempts at implementation of policy.

Resistance to change varies from person to person and they are likely to respond differently to a variety of situations. Change in working practices or environment is another demand increasingly placed upon individuals. With frequent business restructuring, response varies according to the existing demands placed upon the individual and their perceived ability to cope. Stewart (1983) sees these responses as ranging from a temporary disequilibrium to outright rebellion.

Stage 1 Temporary disequilibrium followed by a quick adjustment
Stage 2 Apathy
Stage 3 Antagonism
Stage 4 Open opposition/rebellion/destruction

Within these stages identified by Stewart, individuals faced with frequent alterations in their work situation experience difficulties which influence their behaviour and response to others. The individual arriving at the rebellious phase has experienced tremendous fluctuation in activity that the outrage is likely to lead to arguments with colleagues, sabotage of machinery or engagement in specific practices designed to demonstrate his/her displeasure. Many employees tend to leave the organization before they get to this stage.

Technology has improved a great deal over the last 50 years, particularly in office equipment, e.g. word-processors, electronic mail and fax machines. This can cause uncertainty and fear if employees are not competent to use them. Employees may experience:

- difficulty in adapting, leading to tiredness and resentment;
- increasing job complexity often with no increase in pay;
- a downgrading of work conditions, less mobility and a decrease in social contact at work due to being 'tied' to the screen;
- fears of ill health, e.g. effects of radiation, backache and eyestrain;
- fears of job obsolescence;
- role ambiguities as traditional barriers blur;
- decreasing career prospects if the technology is unfamiliar or not understood, and proper training is not offered.

RECOGNITION AND EFFECTS OF STRESS AT WORK

The effects of stress upon the individual employee within an organization can have negative consequences in terms of:

1. **cost for the employer:** key people leave on the grounds of ill health or as a result of a serious accident;
2. **image:** an organization with a poor reputation regarding the welfare of its employees can lose potential custom and potential employees.

EARLY SIGNS OF STRESS IN THE INDIVIDUAL EMPLOYEE

These include:

- loss of interest in work;
- reduced concentration;
- difficulty in making decisions that were once straightforward;
- decrease in work performance;
- petty theft and vandalism;
- short temper;
- heavier smoking and drinking in those who already do so socially;
- loss of appetite;
- sleeplessness;
- constant tiredness;
- headaches;
- backache;
- indigestion;
- nausea;
- trembling;
- sweating;
- high blood pressure;
- muscle tension;
- blurred vision;
- anger;
- impotence;
- anxiety;
- apathy;
- low self-esteem;
- absenteeism;
- mood swings.

PROLONGED STRESS AT WORK FOR THE INDIVIDUAL

This will result in:

- inefficiency;
- frequent absence;
- social isolation;
- incompetence;
- inability to maintain personal relationships at work.

The results of adverse stress on the individual leads to increased susceptibility to illness. There is considerable debate regarding the precise links between stress and particular physical and mental disorders. One argument suggests that stress suppresses the immune system so that individuals become more susceptible to any type of virus or infection.

Stress is considered to be a significant factor in all of the following disorders.

- Mental breakdown.
- Peptic ulcers.
- Rheumatism.
- Allergies.
- Skin diseases.
- Diabetes.
- Hypertension.
- Heart disease.
- Cancer.

The effects on the organization include:

- reduced quality of service;
- lateness;
- high rates of accidents;
- vandalism and sabotage;
- high numbers of request for early retirement;
- poor working relations, low morale;
- high rates of absenteeism;
- increased disputes and strikes;
- high labour turnover.

MANAGING ORGANIZATIONAL STRESS

Wright (1975) suggested that

> the responsibility for maintaining health should be a reflection of the basic relationship between the individual and the organisation for which he works; it is in the best interests of both parties that reasonable steps are taken to live and work sensibly and not too demandingly.

Managers should become aware of their own stress and the adverse effect it may have on the whole organization. The first step towards this should include the following.

1. ADOPTING GOOD MANAGEMENT TECHNIQUES

- Use good sense and judgement.
- Make the best of generally accepted principles of management.
- Place people in positions for which they are neither under- or over-qualified.

2. CREATING MEANINGFUL JOBS

Commitment is closely related to the individual's perception of the importance of his or her job. Excess stress is reduced when people are committed to their work and have a clear demarcation of roles. Areas of responsibility and authority should be clearly identified; this can be done with the provision of a job description and elements reinforced at induction. Employees should also be made aware of any alterations to their job and be given the opportunity to discuss potential for development of skills or potential alteration in status.

3. MODIFYING THE ORGANIZATION'S STRUCTURE

Managers should ensure that the structure is in keeping with the attainment of organizational goals and that it achieves its ends in a way that causes least disruption.

4. STRENGTHENING PERSONAL QUALITIES

Resistance to stress can be improved by staying in good physical condition, taking positive and constructive action, maintaining interactions with other people and being involved in a creative experience of meaningful work. A purposeful dialogue between managers and workers should be encouraged as it is important for managers to receive and share ideas at all levels of the organization. This assists in maintaining a dynamic and progressive organization.

5. TREAT THE CAUSE OF STRESS

- Assess and adjust distribution of workload.
- Delegate duties and tasks.
- Organize use of time.
- Improve work systems.
- Work with the team not against them.
- Communicate requirements.
- Mobilize systems to eradicate bullying, racism, sexism and any other type of unacceptable, corrosive behaviour within the organization.

Smith *et al.* (1986) note that the most successful attempts to manage stress in organizations take account of:

1. recreating a social, psychological and organizational environment in the workplace to encourage individual autonomy and participation by managers in their jobs;
2. utilizing the social and interactive skills training that help to clarify role and interpersonal relationship difficulties within organizations;

3. creating an organizational climate to foster mutual interpersonal communication, openness and trust among managers and employees.

Managers can assist this process by following Dubrin's (1974) advice in:

- improving communication networks;
- redesigning physical networks;
- managing time;
- taking steps in proactive planning;
- developing career assessment strategies;
- adopting assertiveness techniques;
- developing social-support systems;
- encouraging team-building activities;
- stress management training; and
- fostering good top-down organizational relations.

So far we have looked at the manager's role and personal development in controlling adverse reactions relating to stress in the organization. In the next section, we need to consider the organization and its employees in the prevention of stress-related illness.

ORGANIZATION OF THE WORK ENVIRONMENT

The adverse effects of stress developed within the workplace can lead to accidents and illness. These problems can be reduced by:

- controlling hazards, noise levels, temperature and ventilation;
- provision of well-lit rooms and work stations;
- protecting non-smokers' right to a smoke free environment;
- provision of well-designed and maintained buildings;
- sufficient space to work;
- staff facilities, i.e. dining facilities, coffee lounge, games room, shower, facilities for exercise;
- privacy for staff who need to work without interruptions;
- correct equipment/machinery for the job;
- planned frequent maintenance of equipment/machinery;
- desks/chairs ergonomically sound.

ORGANIZATION OF JOB TASKS

The employee should be provided with sufficient information and resources to complete a product or process. Control of the pace, methods and sequence of work should for the most part be the possession of the employee performing the task.

In addition to this, the job should be designed so that the individuals or teams involved can themselves monitor and be responsible for quality and reliability. There should also be an element of variety, and interest as well as the opportunity for job rotation. Stress can also be reduced if the job allows some social interaction with others.

REVIEW SHIFT ROTAS AND PATTERNS

> A group of female hospital workers were studied in a sample of 1505 women: 43% were nurses, 32% auxiliaries, and 7% ancillary staff; 13% were other qualified health care staff, mainly head nurses; 5% had occupations other than direct health care; 63% worked on the morning, 20% on the afternoon and 17% on the night shift. There is conclusive evidence to state that sleep impairment occurs as a result of shift work patterns and strain.
>
> (Estryn-Behar *et al.*, 1990)

Shifts are less stressful if they are consistent for regular periods of time and do not suddenly change without consultation. There must be long enough break times between shifts to allow recuperation and time for leisure. If shifts do not rotate, it is preferable if the direction of rotation is delayed rather than advanced to give the body's internal mechanisms the best chance to recover. The allocation of shifts should also take account of the employee's personal circumstances, for example, flexible working, job share.

Workload should also be reviewed at regular intervals as well as the employee's training needs.

PARTICIPATION AND CONTROL

> In a study of Australian school teachers it was found that the most prominent stressors in the workplace were concerned with the alienating relationship between teachers and their employers. 96% of those questioned said they lacked influence over decisions which affected them. Lack of influence over organizational decisions and lack of consultation and appreciation by superiors were particular problems.
>
> (Otto, 1980)

Maintaining health in the workplace means that workers and managers need to be able to co-operate and work together in an environment where conflict is controlled, and where participation in the decision-making process is open to all; whatever decisions are made, both managers and workers are likely to be affected by them, so it makes good sense if the decisions taken are mutually beneficial and mutually owned.

Managing stress at work can take the following direction.

CLARIFYING JOB ROLES

- *Provide and update written job descriptions* Ensure every employee is aware of their role within the organization. A full induction programme should be provided for newcomers to the organization. Long-serving employees should be kept abreast of new developments on a regular basis.
- *Improve selection procedures* Equal opportunities legislation must be taken into consideration in all shortlisting of applicants. At interview, candidates feel more at ease if they know that a representative of the panel has received training in gender, racial and disability awareness and is able to represent them fairly.
- *Review job roles* The way work is conducted between different parties, ways this could be changed for the better, what should be kept the same or reduced or increased, who and what needs to change to improve things, how to stipulate goals in such a way that all parties know when they have been achieved.

IMPROVING THE CULTURE OF THE ORGANIZATION

> We found that companies whose only articulated goals were financial did not do nearly as well financially as companies that had broader sets of values.
>
> (Peters and Waterman, 1982)

Managers can improve the company's culture initially through analysing the company's objectives. Each objective should be clearly defined, then communicated throughout the organization using the appropriate channels, i.e. line managers, induction programmes, briefing sessions, email and newsletters.

DEVELOPING AND IMPROVING MANAGEMENT TRAINING TECHNIQUES

Again, managers must be aware that their particular style of managing – or in some cases not managing – has a major influence on the dynamics of the organization. Certain group processes and communications can increase problems and tensions if the organization is poorly managed.

ADOPTING GOOD PROFESSIONAL, PERSONAL AND INDUSTRIAL RELATIONS

- Respect the privacy of others.
- Keep confidences.
- Avoid denigrating employees to their colleagues.
- Encourage use of initiative.
- Put forward and defend own ideas and encourage others to do so.
- Be willing to accept criticism.
- Ask for feedback if it is not forthcoming.
- Use listening skills.
- Negotiate payment systems which reflect the contribution of individuals or groups of employees.
- Discuss the implications of changes in tasks, work organization and endeavour to resolve conflicts quickly.
- Consider any special needs that employees may have and make provisions to assist them in making appropriate arrangements, e.g. maternity, bereavement and other kinds of special leave.
- Take steps to deal with sexism, racism and other forms of prejudice.

PROMOTING HEALTH IN THE WORKPLACE

> if you want productivity and the financial rewards that go with it, you must treat your workers as your most important asset.
>
> (Peters and Waterman, 1982)

Managers are responsible for the health, safety and well-being of employees in the work setting; it therefore makes sense if they invest in their health. Promoting health at work is noted to have many benefits. These include:

- better health for employees;
- improved morale;
- reduced stress;
- increased job satisfaction;
- better working environment;
- reduced absenteeism;
- higher productivity;
- better industrial relations;

- higher rates of productivity;
- fostering a positive corporate image.

The first step towards managing stress in any organization should be the drawing up of an action plan.

MANAGEMENT ACTION PLAN

1. FORM AN ACTION COMMITTEE

This can be made up of representatives throughout the organization, e.g. line managers, health and safety officers, employees, equal opportunities representatives, union officials, occupational health specialists, educationist, marketing officer, catering staff.

2. ASSESS THE SITUATION

Identify the possible causes of stress in the organization. This may be discovered by discussions with workers, noting production levels, staff turnover rates, sickness/absence records, observation of staff morale, conducting research. Examine current practices.

3. CONSIDER FINDINGS AND PRIORITIES

Are there any common patterns emerging from the evidence collected?

4. FORMULATE A POLICY TO REDUCE STRESS WITHIN THE ORGANIZATION

- How can the organization solve the problem? Is there anything that needs to be addressed urgently?
- Work out a strategy for achieving aims.
- Are any costs likely to be incurred? Can work be carried out in-house? Do experts need to be consulted?
- Set out clear aims and objectives.
- Formulate and write up policy that is suitable to the majority.

5. IMPLEMENT POLICY

Raise awareness by using line managers who can ensure that each member of the organization has access to a copy of the policy. Appoint individuals with clear roles to implement changes and put support systems in place.

6. EVALUATE POLICY AND INITIATIVES

The action committee should modify and revise initiatives and policy as necessary, taking note of progress reports, complaints and spin-off activities.

SUMMARY ACTION PLAN

- Form action committee
- Assess situation
- Consider priorities
- Formulate policy
- Implement policy
- Evaluate

GUIDANCE FOR MANAGING STRESS AT WORK

RAISE AWARENESS

Provide knowledge of services/initiatives available using the most effective means of communication possible throughout the organization.

PROVIDE ACCESS TO SERVICES

For example:

- occupational health nurse/doctor either on site or close by;
- marital guidance and counselling services;
- relaxation classes;
- stress-management courses;
- sport and leisure facilities;
- time-management courses;
- t'ai chi;
- meditation classes;
- yoga.

PROVIDE FEEDBACK MECHANISMS FOR WORKERS

Employees are the best people to supply a variety of viewpoints in order to develop, modify and extend services within the organization.

INDIVIDUAL COPING STRATEGIES

Employees will have different ways of coping with problems arising from stress. Some examples include:

- treating oneself occasionally;
- talking to a friend;
- developing and using support networks;
- taking alcohol in moderation;
- sensible diet; a balanced diet influences mood, provides energy and healing;
- control weight;
- reduce smoking;
- adequate rest and sleep;
- pace work;
- unwinding after work, e.g .warm bath, walking the dog, listening to music, reading, meditation;
- taking exercise; this promotes mental well-being, increases flexibility, strength and endurance. The exercise regime should be tailored to the individual's needs.

WORK MANAGEMENT

Managing stress at work will also involve:

- varying the work routine;
- taking breaks when appropriate;
- making full use of holiday allocation;
- taking full lunch breaks;
- developing a life outside work;
- setting priorities;
- managing time effectively;
- establishing a manageable work rate and sticking to it;
- delegation;
- knowing when to say no;
- managing conflict sensitively; knowing when to withdraw strategically;
- being assertive.

COPING WITH ORGANIZATIONAL STRESS

Managers may take note of the Norwegian methods for combating unwarranted stress in the organization. For over a decade, employees have worked in accordance with the legislation appertaining to the workplace and health.

The Norwegian Working Environment Act requires that technology, work organization, work times, e.g. shift plans and payment systems, are to be so designed as to enable the avoidance or at least reduction of negative psychological or physiological effects for employees. Employees are also to be given the opportunity for personal development and for the maintenance and development of their skills.

Recommendations have also been made by experts in the field with regard to the design and planning of job tasks, and possibilities for employee self-determination and maintenance of skills. Monotonous repetitive work, and work that is bound by machinery or assembly line in such a way that there is no room left for variation in work rhythm, should be avoided. Jobs have been so designed to provide possibilities for variation, for contact with others, for understanding of the interdependence between elements that constitute a job and for information and feedback to employees concerning production requirements and results. Workers and their elected representatives are kept informed about systems used for planning and control and any changes in such systems. They are given the training necessary to understand such systems and influence their design.

In addition to the Norwegian example, other researchers in the field have made further recommendations. Smith *et al.* (1986) recommend the following interventions in managing stress within the organization.

- To recreate a social, psychological and organizational environment in the workplace to encourage individual autonomy and participation in planning their jobs, so achieving fulfilment.
- To utilize the social and interactive skills training that help clarify role and interpersonal relationship difficulties within the organization.
- To recreate an organizational climate to foster mutual interpersonal communication, openness and trust among managers and employees.

Prevention and change at organizational level is more efficient than teaching large groups of people to cope with excessively high stress levels or undertaking some therapeutic intervention. Employee uncertainty, ambiguity and feelings of helplessness can be lowered by reducing environmental stressors. This can be accomplished by job redesign and job enrichment, clarifying performance evaluation standards, decentralizing authority where appropriate, and improving communication networks. A redesign of physical surroundings can reduce stress. Education/training in problem-solving techniques such as time management, proactive planning, career assessment and assertiveness techniques can also lessen dysfunctional behaviours related to stress (Mallinger, 1986). Social support and team building activities can also play a role in decreasing organizational stress. Finally, stress management training can be instituted either commercially by contracting with private employee assistance programmes or internally through corporate training programmes.

Any stress management programme that is introduced in a hurry and without employee support will be unsuccessful. There must be a high level of motivation and participation. This can be achieved by getting employees involved, e.g. by telling their colleagues, compiling lists for training and appropriate advertising within the organization. Top-down organizational support will complement and reinforce the commitment to behavioural change (Kindler, 1984).

Effective stress management should develop employees' awareness of stress and how it affects their lives both positively and negatively. Skill acquisition will assist in identifying specific techniques employees can initiate to prevent negative stress reactions or to dissipate these reactions more productively. Feedback provides participation with specific information relating to their current level of stress, certain behavioural patterns and personal orientation (Ivancevich and Matteson, 1982).

Managers should also make use of consultants/experts in the field to identify the sources of stress within the organization. The Occupational Stress Indicator (OSI), developed by Cooper, Sloan and Williams in 1988, is now one of the most widely used tools for detecting stress-related problems at work. The analytical tool comprises 167 questions relating to aspects of general behaviour, control and influence within the job, methods of coping with stress and so forth, which take up to 45 minutes to complete using a computer terminal. The scores are adapted to relate to both professional and manual worker groups. Cable & Wireless (UK) have adopted this system and have designed work programmes specific to the needs of the employee and the organization with considerable success.

SUMMARY

Unwarranted stress occurs as a result of interaction in a variety of situations in the workplace as well as poor stimulation, poor motivation, monotony and overwork. Structural problems that encourage prejudice in all its forms is also responsible for this adverse reaction among the workforce. Managers can assist in its reduction by adopting various strategies to benefit the organization and the individual employee. These include:

- action planning;
- consideration for others;
- provision of sufficient training to do the job;

- providing specific and clearly written job descriptions;
- establishing good communication systems;
- monitoring the causes of sickness absence;
- evaluation of the organization's goals.

SUGGESTED FURTHER READING

Cooper, C.L. and Marshall, J. (1980) *White Collar and Professional Stress*, Chichester and New York. Wiley.

Cervinka, R. (1993) Night shift and stress at work, *Ergonomics*, Jan.–Mar., **36**(1–3), 155–60.

Mallinger, M. (1986) Stress management: an organizational and individual responsibility, *Training and Development Journal*, Feb.

REFERENCES

Beehr and Newman (1978) Job stress, employee health and organizational effectiveness. A facet analysis model and literature review, *Personnel Psychology*, **31**, 665–99.

Cervinka, R. (1993) Night shift and stress at work, *Ergonomics*, Jan.–Mar., **36**(1–3), 155–60.

Clarke, T.A., Maniscalco, W.M., Taylor-Brown, S., Roghmann, K.J., Shapiro, D.L. and Hannon-Johnson, C. (1984) *Paediatrics*, **4**(1), July, 52–7.

Cooper, C.L., Sloan, S.J. and Williams, S. (1988) *Occupational Stress Indicator: Management Guide*, NFER-Nelson, Windsor.

Dubrin, A. (1974) *Fundamentals of Organizational Behaviour*, New York. Pergamon Press.

Estryn-Behar, M., Kaminski, M., Peigne, E., Bonnet, N., Vaichere, E. and Gozlan, C. (1990) *British Journal of Industrial Medicine*, **47**(1), Jan., 20–8.

Gardell, B. (1977) Autonomy and participation in work, *Human Relations*, **30**(6), 515–33.

Gardell, B., Aronsson, G. and Barklof, K. (1982) *The Working Environment of Local Transport Personnel*, Swedish Work Environment Fund, Stockholm.

Ivancevich, J. and Matteson, M. (1982) The how, what and why of stress management training, *Personnel Journal*, Oct., 772.

Kindler, H. (1984) Time out for stress management training, *Training and Development Journal*, June.

Locker, D., Burman, D. and Otchere, D. (1989) *Community Dental-Oral Epidemiology*, **17**(5), Oct., 263–6.

Mallinger, M. (1986) Stress management: an organizational and individual responsibility, *Training and Development Journal*, Feb.

McGrath, J. (1976) Stress behaviour and organisations. In M. Dunnett (ed.), *Handbook of Industrial and Organisational Psychology*, Chicago: Rand McNally.

Otto, R. (1980) *Occupational Stress among Factory Workers*, La Trobe University Working Papers in Sociology, 58, Bundoora.

Peters, T. and Waterman, R. (1982) *In Search of Excellence: Lessons From America's Best Run Companies*, Harper & Row.

Seyle, H. (1978) *The Stress of Life*, New York, McGraw-Hill.

Smith, M., Beck, J., Cooper, C. L., Cox, C., Ottaway, D. and Talbot, R. (1986) *Introducing Organisational Behaviour*, London, Macmillan.

Stewart, V. (1983) *Change: Challenge For Management*, New York, McGraw-Hill.

Webb, T., Schilling, R., Jacobson, B. and Babb, P. (1989) *Health at Work? A Report on Health Promotion in the Workplace. Research Report* **22**(24).

Wright, H. (1975) *Executive Ease and Dis-ease*, Epping, Gower Press.

Zalenznik, A. (1977) Stress Reactions in Organisations, *Behavioural Science*, **22**(3), 151–162.

USEFUL ADDRESSES

International Stress Management Association (ISMA) UK, 156 Westcombe Hill, London SE3 7DH.

Relaxation For Living, 160–170 Oatlands Drive, Weybridge, Surrey KT13 9ET.

British Association For Autogenic Training and Therapy (BAFATT), 101 Harley Street, London W1N 1DF.

British Association For Counselling, 37a Sheep Street, Rugby, Warwickshire CV21 3BX.

Managers' Counselling Association, Sunridge Park Management College, Plaistow Road, Bromley, Kent BR1 3PT.

Health and Safety Executive, Library and Information Services, Broad Lane, Sheffield S3 7HQ.

Back pain and back injuries 3

Back problems at work • The back • Patterns of occurrence and recurrence • Potential causes of back pain • General causes of back pain • Back problems among nursing personnel • Managing back problems at work • Back injury prevention among truck and delivery van drivers • Summary • Suggested further reading • References • Useful addresses

BACK PROBLEMS AT WORK

The workplace is the single most common factor in contributing to back injury as studies of back pain and back injuries at work show. More than a quarter of the accidents reported each year to the enforcing authorities in the UK alone are associated with manual handling, that is the transporting or supporting of loads by hand or bodily force. Sickness absence from back pain is increasing.

- Back pain currently costs the UK about £3 billion in lost output each year.
- Each working day 230 000 people in the UK are off work with back pain.
- The cost of treating back pain in the NHS is over £350 million each year.
- At least 80% of men and women will suffer some form of back pain during their lifetime.
- Back pain at work means:
  - reduced performance;
  - increased sickness absence;
  - increased sick pay;
  - loss of trained and experienced staff;
  - increased insurance premium.

The vast majority of reported manual handling accidents in the UK result in injury causing absence for more than three days, most commonly a sprain or strain of the back. Workplace occurrences fall into three categories:

- **task demands;**
- **lack of control over the work situation;**
- **positioning.**

1. TASK DEMANDS

In the early 1970s, Margora published a series of important studies concerning the effects of occupational task demands on the prevalence of low back pain (Margora, 1970a, 1970b, 1972, 1973a, 1973b). He studied more than 3000 Israeli men and women from eight different occupations. According to the data, people who frequently lift weights are likely to show a high prevalence of back pain.

Kelsey *et al.* (1984) also found that jobs involving the lifting of objects which weighed more than 25 lb (11.3 kg) carried a threefold increase of risk. This risk was extended if twisting was involved, particularly if the knees were straight.

Frymoyer *et al.* (1980, 1983) confirmed that jobs involving lifting, carrying, pulling, pushing, bending and twisting all carried significantly increased risks; also, an investigation of women in the retail trade found an association between back pain and the need to perform frequent (i.e. more than 10 times per hour) twisting and reaching movements.

2. LACK OF CONTROL OVER THE WORK SITUATION

Vallfors (1985) confirmed that people with back problems are significantly less likely to have varied working tasks or the ability to choose their own working position.

Margora's studies also found an association between back pain and low job satisfaction. This may be partly because people who do repetitive jobs in fixed working postures are more likely to have low job satisfaction. Subsequent studies have found that people with back problems are more likely to report stressful events at work and have a lower opinion of the working environment as a result.

3. POSITIONING

Wickstrom (1978) and Buckle's (1983) studies of foundrymen and turners (lathe operators) found that, on average, men with back problems spend longer in a stooped position than on controls. This places tremendous strain on the back and neck muscles.

We shall return to the problems associated with the workplace later. First, a look at the back and its functions.

THE BACK

The back extends from the neck to the coccyx and it is the region between the neck and lumbar region where most difficulties tend to arise.

The functions of the back are:

- to provide support to the body as a whole;
- to provide protection for the spinal cord;
- movement: the spines on the vertebrae provide surfaces for the attachment of back muscles.

The back bone, known as the vertebral column, consists of 33 irregular bones which are divided into the following groups.

MOBILE VERTEBRAE

The moveable parts of the back consist of:

- 7 cervical vertebrae in the neck; these are the smallest of all because they carry least weight;
- 12 thoracic or dorsal vertebrae forming part of the posterior thoracic wall; these possess flat surfaces for articulation with the ribs;
- 5 lumbar vertebrae forming part of the posterior abdominal wall. These are the largest bones and they carry most of the body's weight.

IMMOBILE VERTEBRAE

The immobile vertebrae consists of:

- 5 sacral vertebrae fused together to form a bone called the sacrum which is part of the pelvic girdle;
- 4 coccygeal vertebrae fused together to form the coccyx or tail remnant, which is part of the pelvic girdle.

Between the mobile vertebrae lie discs, which are pads of white fibrocartilage. Their functions are:

1. to protect the brain and spinal cord from injury by acting as shock absorbers;
2. to allow a limited amount of movement of the vertebral column.

In the neck and lumbar region there are many structures where problems can develop:

1. the vertebral bones can change shape with a chronic disease such as arthritis, and
2. there are many muscles attached to these bones. These are directly attached by ligaments which may themselves become diseased, worn or torn.

In addition, the spinal cord is protected by the bony vertebrae but the nerves from the spinal cord pass through and between the muscles, ligaments and bones. They can be diseased themselves or may be affected by the diseased or altered bones, ligaments and joints between which they pass. The best known is the slipped disc where the disc protrudes between two vertebrae and presses on a nearby nerve as a result of dislocation. Pain arises which can be excruciating.

Low back pain is the most common of the work related musculoskeletal disorders and in economic terms the most costly. Pain in the lower back may be sharp or dull, diffuse or localized. When the pain is acute in onset or severity, it is sometimes called lumbago. The pain may be limited to the midline or it may extend outward on either side and, in around one-third of cases, it radiates into the buttocks and down the back of the thigh and into the leg (sciatica). This may be associated with tingling in the foot and other signs which may be suggestive of nerve root involvement. The pain may also radiate into the groin and down the front of the thigh.

Muscle spasm and limitation of movement are common findings. In some severe cases they become more pronounced.

Garg and Moore's study (1992) noted jobs involving lifting, lowering, pushing, pulling, carrying and holding; body movements such as frequent bending, twisting and sudden movements, as well as working in bent over postures, appear to have significant potential for producing low back pain. A combination of lifting, bending and twisting appears to be the most hazardous. It is concluded that lifting heavy loads contributes to increased frequency and severity rates for low back pain. This is true regardless of whether the lifting is performed over a short period or throughout the day and whether it is performed at least three times a day or repetitively. If, however, such lifting is performed repetitively, the medical hazard extends beyond low back problems to other musculoskeletal strain and sprain injuries and to fatigue-related injuries, particularly in those who are less strong.

PATTERNS OF OCCURRENCE AND RECURRENCE

Most episodes of low back pain are relatively short-lived. The symptoms usually subside within a few weeks at most although some episodes may be prolonged. The rate of recurrence is very high.

Various studies have shown (Beiring-Sorensen, 1983; Troup *et al.*, 1981) that of patients who had recovered from an attack of back pain, it was found that between 50% and 60% had another attack within a year. Those individuals previously experiencing back troubles are never quite symptom-free. They are likely to experience dull aches periodically, made worse by certain activities and punctuated by episodes of intense pain.

> A study carried out in Glostrup, Copenhagen, investigated back pain prevalence amongst men and women aged between 30 and 60 years. The overall point of prevalence for 'pain or other trouble in the lower part of the back' was 14%; the one year prevalence was 45% and the lifetime prevalence was 62%.
>
> (Beiring-Sorensen, 1982)

Trends in rates of low back strains, low back impact injuries and non-low back injuries among field employees of a petroleum drilling company between 1979 and 1985 were examined to identify the relationship between economic factors and the incidence of low back pain and other injuries. Economic indicators included the rate of resignations, a surrogate for turnover and the rate of layoffs. Only time lost due to low back strain increased during times of worker layoffs (Clemmer and Mohr, 1991).

POTENTIAL CAUSES OF BACK PAIN

There is a lack of clarity concerning the cause and effect relationships between identifiable and quantifiable risk factors in low back pain. However, a variety of risk factors has been identified such as:

- family history;
- clinical history of low back pain;
- height;
- body mass;
- psychosomatic fear;
- intelligence;
- level of education;
- metabolic cost;

- cigarette smoking and chronic coughing;
- strength;
- pregnancy;
- prolonged sitting.

Evidently, many of these lie outside the influence of the employer. However, using epidemiology as an information base and task design as a potential strategy, Snook (1978) asserted that up to one-third of back injuries can be prevented. It was found that up to a quarter of jobs required manual handling of a magnitude acceptable to less than 75% of the workers. These jobs were responsible for the back injuries reported; indicating a threefold increase in susceptibility when the tasks required much exertion. This implies that two-thirds of all back injuries associated with manual materials handling can be prevented if the job is designed to fit at least 75% of the workforce. The survey discovered that one-third of the injuries will occur despite any design effort. Counting the injuries which were not associated with heavy manual handling, Snook calculated an overall possible reduction in low back injuries by 33% for industrial workers if an appropriate design strategy were to be instituted. Though such claims may be difficult to substantiate, it indicates an approximate 33% collective margin of safety for the workforce.

GENERAL CAUSES OF BACK PAIN

The general causes of back pain are:

- normal wear and tear as a result of the ageing process;
- poor physical fitness;
- bad working habits;
- stress;
- muscular tension;
- poor posture;
- loss of control on lifting load;
- twisting trunk leading to stress on the lower back;
- stooping;
- reaching upwards leading to stress on the arms and back;
- carrying a load for an excessive distance, i.e. if a load is carried further than about 10 metres then the physical demands of carrying the load will tend to predominate and the individual's capability will be reduced;
- insufficient rest/recovery period after carrying one load then another.

ACUTE BACK PAIN

Acute backache (lumbago) may be an indicator of a specific condition such as prolapse of an intervertebral disc. Sometimes a fringe of synovial membrane is nipped between the articular surfaces of one of the small intervertebral joints. The result is sudden pain precipitating the acute muscle spasm which is the dominant feature of lumbago.

Alternatively, a few fibres of one of the spinal ligaments may be torn, leading to the same chain of events. Whatever the cause, acute lumbago is commoner in individuals who have either some minor structural anomaly or muscular insufficiency.

Backache of sudden onset in the lumbar region is usually caused by a sudden unguarded movement. The pain may be in the midline or to one side usually in the sacro-iliac region. It is aggravated by any movement and relieved by rest, usually lying flat on the back. Sitting tends to increase the discomfort, but when the pain is in the upper lumbar region, it may afford relief. If there is radiation away from the back, this indicates irritation of a nerve root at the site of the causal legion, suggesting a prolapse of an intervertebral disc as the cause of the condition.

CHRONIC BACKACHE

This commonly follows an episode of acute backache, which after an initial improvement then remains as a nagging ache, worse after any strenuous activity, particularly if this involves stooping or lifting. It may affect any level of the back, but because stresses are greatest in the lumbo-sacral region, this is the most frequent site. Chronic backache may also be the result of mechanical instability of the lumbar spine. The facet joints are believed to undergo degenerative changes as occur elsewhere in the body, and pain develops from increased movement at these joints.

BACK PROBLEMS AMONG NURSING PERSONNEL

Manning *et al.* (1984) investigated events associated with the onset of episodes of back pain in a working population. These were divided into accidents *per se* and what they referred to as 'non-accidental injuries'. Non-accidental injuries are injuries occurring in the course of the individual's normal working activities, without the intervention of any unforeseen event other than the pain itself. The accidents and non-accidental injuries accounted for 30% of cases each and in about 40% of cases the condition developed insidiously. The most common accident was loss of footing (66%), followed by an unexpected load (12%) or loss of balance (10%). Non-accidental injuries most commonly occurred during bending/straightening actions (73%), lifting and carrying (50%) and twisting (48%).

Back injuries have been a particular problem in the nursing profession as a whole.

In the USA, approximately one in ten compensable medical cost is for low back pain. Medical care has been estimated at over $£ billion. Over-exertion has also been associated with manual materials handling jobs, accounting for 12 million lost working days.

Nursing personnel have high prevalence rates of back pain as well as high incidence rates of workers' compensation claims for back injuries. Researchers have indicated that the low back pain experienced by nurses is greater than the published statistics indicate and nurses perceive back pain as an inevitable part of nursing practice (Kelsey, 1982; Klein *et al.*, 1984; Beiring-Sorensen,1985; Harber *et al.*, 1985; Jensen, 1987; Personick, 1990; Garg *et al.*, 1991).

Male careworkers have also filed between 77% and 80% of the back injury compensation claims in 26 American states. The highest claims were made by men aged 20–24 years, compared to female claim rates being in the 24–34 year age category. For both groups, the highest risk of compensable low back strain occurred between ages 20–44 years.

In the UK, back injuries too are a major cause of sickness absence among nurses. At least one in four nurses reported having time off with back injury sustained at work (IMS, 1992). These injuries have led to long-term redeployment and premature retirement. Indeed, the loss of trained nurses as a result currently costs the NHS £50 million a year.

Back-injury problems appear to be even greater among careworkers in nursing homes than in hospitals. This is increasing. Several studies have concluded that frequent manual lifting and transferring of patients are the primary, or at least the most recognized, causal factor for low back pain. In addition to this, many nurses believe that of all the tasks performed by them, patient handling activities are the most likely to result in low back pain (Bell *et al.*, 1979; Dehlin *et al.*, 1976; Greenwood, 1986; Harber *et al.*, 1985; Jensen, 1990; Personick, 1990).

Many injury prevention programmes among nursing personnel tend to focus on sound lifting techniques, body mechanics and body care. A popular view among this occupational group is that the main way to avoid back injuries from patient handling activities is always to apply proper body mechanics. But the fact is that some patient handling tasks are so stressful that back injuries, despite taking all the correct precautions, are not amenable to the use of proper body mechanics. In addition, the current consensus appears to be that concentrating on the individual and the technique alone does not prepare the nurse for every eventuality. The ergonomic approach – adjusting work design, tasks and equipment – is suggested as most effective in preventing stress, strain and accidents among nurses.

MANAGING BACK PROBLEMS AT WORK

A number of ways have been suggested to assist managers in helping themselves and their staff to prevent back pain and injuries in the workplace. When planning to modify any aspect of workplace behaviour, a strategic management tool is considered useful.

Van Akkerveeken (1985) identifies the necessity for strategic planning and communication among managers and their team as influential factors in improving conditions in the workplace. These factors include:

- workers sharing the goals and values of the whole organization;
- a chain of communication between workers, supervisors, management, insurers, unions, physicians and therapists;
- the active involvement of managers in the formulation of strategies for prevention.

Any alterations in the values or the culture of the organization will need consideration and the use of a tried and tested model. Urlings *et al.* (1990) suggest a six-stage model for attitude and behaviour change in management, based on psychosocial and communication theories already adopted in the Dutch health education system (see Table 3.1). Analytical tools within the model are used continually to discover problems within the process so that efforts are usefully concentrated.

Table 3.1 Six-stage model for attitude and behaviour change in management

Stage	Process	Influential factor
1	Attracting attention for the offered information	Selective perception of the offered information
2	Understanding of the offered information	Understandability of the offered information
3	Changing the attitude	Expected advantages and disadvantages of the new behaviour
4	Changing the intention	Expected social pressure or social support
5	Changing the behaviour	Possibilities and impossibilities of performing the behaviour
6	Maintenance of new behaviour	Feedback about the new behaviour

(Source: Urlings *et al.*, 1990.)

The cultural change within the organization should also include the sharing of information and knowledge. A two-way communication process should be adopted to identify the needs of the employee and manager.

A useful assessment of the needs of employees can only be based on a thorough and practical understanding of the type of manual handling tasks to be performed. This includes knowledge of the loads to be handled and the working environment in which the tasks will be carried out. Shopfloor managers and the workforce are better placed to know about the manual handling taking place in their organization than any external source.

More specifically, assessment should include observation of the operations that workers are likely to perform. Question whether the operations are potentially hazardous and if they can be avoided or mechanized or automated at reasonable cost. Also note whether the operations are within the current guidance provided by health and safety experts.

Do any of the operations involve:

- holding loads away from the body?
- twisting, reaching upwards or stooping?
- any large vertical movements, long carriage distance or unpredictable movements of load?
- strenuous pulling or pushing?
- repetitive handling, insufficient rest or recovery?

Load will be another factor in the assessment of operations to prevent back injury. The assessor should note whether any objects will be difficult to grasp, are unstable, heavy, bulky or are likely to cause harm if lifted without mechanical or other human assistance. Along with assessing load, the individual(s) involved in any lifting operation must be capable, be physically fit and aware of their own capabilities.

As noted in the extract on the top of the facing page, the work environment is an important consideration in assessing hazard. When looking at potential causes of back strain or back injury, does the work area:

- constrain posture?
- have poor flooring or variation levels?
- have fluctuating and inconsistent changes in temperature?
- have poor lighting?

Are any of these a risk to the individual and what action can be taken to reduce any compensation claims against the employer for negligence? Here are a few suggestions.

> The management of a large UK warehousing and distribution operation wanted to know the maximum safe weight of supplies cages handling by the distribution drivers in order to minimise the risk of back injury. A study was undertaken in which the force applied at the hands and the load at the lower back were measured, whilst cages loaded to known weights were manoeuvred according to set procedure.
>
> The data was compared with recommended maximum pushing and pulling forces in order to provide the guidance requested by management. A number of reference sources were considered, including psychophysical data and the 'force limit' data, but a lack of consensus was found. In addition, hand height, position of feet relative to the load and the coefficient of friction between the shoe and the floor have a major bearing on the magnitude of acceptable pushing and pulling forces. Although training could be of assistance in this regard, the drivers and managers had little control over the environment in the customer's premises through which the loaded cages have to be transported. So although it was possible to advise management on an appropriate loaded cage weight, it was necessary to stress the limitations within such a working environment.
>
> (Buckle *et al.*, 1992)

INVOLVE THE WORKFORCE

Employees and health and safety representatives should be involved in any discussion and plans for redesign or reorganization of the system of work and be encouraged to report its effects once the new system is in place. Opportunity should be given to contribute to the development of good handling practice.

MONITORING

Monitoring effectiveness of any new developments either through general surveys within the organization or directly questioning the users of any new machinery, or recipients of training for the purpose of manual handling.

IMPROVE TASK LAYOUT

Improving flow of materials or products is likely to increase productivity and efficiency. Materials that are heaviest should be stored at waist height. Lighter or more easily handled loads should be stored above or below waist height.

USE THE BODY MORE EFFECTIVELY

When lifting heavy items or equipment, employees should be advised to adopt a method that utilizes a series of operations in sequence or one that reduces the need for stooping, twisting or stretching. The load should also be held as close as possible to the body to reduce any risk of injury.

> Regulation 12 of the Management of the Health and Safety at Work Regulations 1992 (EC) requires employees generally to make use of appropriate equipment provided for them, in accordance with their training and the instructions their employer has given them. Such equipment will include machinery and other aids provided for the safe handling of loads.

Lifting loads that are near to or at floor level is sometimes unavoidable. In such circumstances, handling techniques which involve the use of the stronger leg muscles rather than those of the back are preferable.

WORK ROUTINE

Periods of strenuous work activity may if possible be interrupted with lighter activities such as paperwork. Adequate and regular rest breaks must also be encouraged if the daily routine involves the frequent lifting of loads. Job rotation can also bring reduced monotony and increased attentiveness.

Team handling will help to lighten the load; the individual team members in an ideal situation should be similar in height and physical capability. Access to handling aids will also be useful as well as protective gloves, aprons, overalls and sensible footwear.

ATTEND TO THE WORK ENVIRONMENT

Managers must ensure that their workforce feel safe and well equipped to carry out their tasks. In manual handling situations ensure the following.

- Sufficient floor space that is clean and grease-free.
- Ample head room to carry out manoeuvres.

- Well-ventilated and properly drained surfaces outdoors.
- Any temporary work platforms should be firm and stable.
- Spillages to be cleared up immediately.
- Avoid lifting objects from sloped surfaces.
- Provide sufficient and well-directed light to enable handlers to carry out operations sufficiently.
- Attention to work surface design.

A redevelopment of baggage handling facilities at Heathrow Airport, London (UK), had led to concern among safety, management and union personnel that a potential back injury hazard had been created. The cause of concern lay in the design of the newly installed conveyor system. This new system was also to be the design for all future baggage handling lines. A vertical lift of 53 cm from the base of the incoming baggage container was necessary to place the baggage on the conveyor. With the old system, the lift distance was only 4 cm. Management wanted to know the following:

(I) whether handling baggage from containers on the new conveyor was likely to lead to back injury for the operatives.
(II) whether the existing design was preferable in this respect.

On evaluation, the force limits data were found to be inapplicable. They placed emphasis on the weight of the load and not on the lifting distance.

As a consequence, management decided to revert to the existing system, even though evidence from the study questioned the effectiveness of the design from an injury prevention perspective.

(Nicholson and Clark, 1986)

PROVIDE TRAINING AND INFORMATION

Section 2 of the Health and Safety at Work Act 1974 in Britain and Regulations 8 and 11 of the European Commission's Management of Health and Safety at Work Regulations 1992 require employers to provide their employees with health and safety information and training. Employers should ensure that their employees understand clearly how manual handling operations have been designed to ensure their safety. Employees, their safety representatives and safety committees should be involved in the monitoring of its effectiveness.

Devising a training programme for safe lifting and back protection should involve consideration of the following.

- Recognition of potentially hazardous handling operations.
- How to deal with routine and unfamiliar handling operations.
- The best use of handling aids.
- The use of personal protective equipment.
- Features of the working environment that contribute to its safety.
- Factors affecting individual capability.
- Action that may be necessary in an emergency.
- Good handling technique.

> Managers should be aware that Section 7 of the Health and Safety at Work Act 1974 (UK) states that employees must:
>
> take reasonable care for their own health and safety and that of others who may be affected by their activities. They should also co-operate with their employers to enable them to comply with their health and safety duties.

APPLICATION OF ERGONOMICS

Back injury prevention not only requires education and training of individuals within an organization, it also involves some application of ergonomics. Instead of focusing on the behaviour of people, the ergonomic approach involves designing or readjusting work and workload so the physical and mental demands of tasks to be undertaken are within the capabilities of the worker.

This approach has been widely recommended (Bell, 1984; Brown 1972; Buckle 1983; Dehlin *et al.*, 1976; Garg *et al.*, 1991; Harber *et al.*, 1985; Jenson, 1990; Lloyd *et al.*, 1987; Snook *et al.*, 1978; Stubbs *et al.*, 1986; Videmean *et al.*, 1984). In addition, it has been suggested that application of an ergonomic approach to patient care jobs could lead to safer and more efficient methods of handling patients which will reduce certain levels of low back stress particularly among nurses.

Ergonomics have been useful in altering job tasks and the use of specially designed equipment to aid lifting. For example, Snook (1978) found that changing the size and shape of containers in his study was found to be easier and safer to lift, and reduced the risk of back injury. Wheeled containers for sanitation workers, 'arms' and wheels on garbage containers were also found to reduce the stress of lifting.

MONITORING

Regular review of sickness absence is necessary to discover the frequency with which workers take sick leave as a result of back injury.

Note if the causes are work-related and match this finding with any accidents reported involving the back. This information will be useful in any training or awareness-raising initiative undertaken within the organization. It is also vital information for identifying the cost of sickness absence as a result of back injury on a regular basis.

> Analyses of WorkCare data from the State of Victoria in Australia revealed that claimants with back injury claims of greater than 12 months standing were very costly. A total of 8633 claimants accounted for 68.7% of all payments for back injury ($362 million out of a total of $527 million) when these claimants represented only 10.7% of all back injury claimants for the period September 1985 to May 1989.
>
> (Thomas *et al.*, 1990)

RAISE AWARENESS

Preventing back injury is the responsibility of the whole organization not merely that of the conscientious manager or considerate employee. Thus any awareness-raising strategies should employ information that is relevant to all. Such information should include:

- knowledge of the body and its limitations;
- safe systems of working;
- location of the nearest occupational health department and/or access to treatment;
- reporting and recording of incidents;
- how to lift and handle materials correctly.

The format of raising awareness can include briefing sessions, training, newsletters, leaflets and posters. Health education/health promotion units also provide literature that raises awareness and informs individuals to look after their back. Ensuring the whole organization is aware of back injury prevention will go some way to maintaining compliance with the law.

UTILIZING SPECIALISTS

1. Physiotherapists

A physiotherapist manipulates bones and realigns joints, massages muscles to relieve spasm and tension, and provides instruction and advice on appropriate exercises to improve or maintain physical condition. Electrotherapy may also be employed if pain relief or the promotion of healing is required in the event of strain or injury. Physiotherapists will also act as useful guides in training in manual handling techniques.

2. Chiropractitioners and osteopaths

They specialize in manually treating musculoskeletal conditions and aim to restore normal structure of the spine by relieving pressure on the nerves. They provide assistance and advice on maintaining a healthy posture.

3. Occupational therapists

Occupational therapists work with people who have physical or mental illness, disability or severe injury. This enables those who are temporarily or permanently disabled as a result of work-related injury to be as independent as possible in everyday activities. Treatment includes adapting or providing equipment or recommending alterations to buildings, providing suitable activities to help improve abilities affected by illness or accident and enable people to practise everyday tasks.

BACK INJURY PREVENTION AMONG TRUCK AND DELIVERY VAN DRIVERS

The following case is an example where normal training techniques could only be employed in a limited sense, and where it was discovered that the application of ergonomic methods was best employed.

Injuries among drivers of delivery vans and trucks have been increasing within the last ten years. This has lead to a rise in worker compensation claims particularly in the soft drinks industry (Walker, 1985). There have been several explanations offered for this occurrence:

1. market competition and the increased handling of goods in the industry;
2. diversity in soft drink packaging and inconsistency in load;
3. larger trucks with more products to be delivered.

Despite the changes in the industry, the driver is still required to perform the duties of driving, delivering and unloading the soft drink products.

Information gathered by the Division of Physical Science and Engineering Survey, concluded that no single engineering and/or administrative control would sufficiently reduce the manual handling problems experienced in the soft drinks industry. Instead, a series of control measures were recommended. They include:

- reduction in the amount of handling, sorting and rehandling of beverage products;
- use of electric forklift trucks at the site of delivery to unload and transport pallets of beverage containers;
- provision of handholds in all boxes – standardize the size and location of handholds for consistency;
- developing a labelling system on boxes so product identification is unmistakable;
- installing access ramps to kerbs and stairs.

These measures were considered the best way of limiting difficulties relating to manual handling. It will be interesting to see how the industry has faired in years to come.

SUMMARY

In any initiative to prevent back problems and back injuries in the workplace, the following suggestions may act as a useful guide.

- Assess any potential hazards – involve the workforce.
- Adhere to most current manual handling regulations.
- Update workforce on most current legislation and their implications.
- Provide training and updated equipment for use by employees.
- Check sickness and absenteeism records on a regular basis – note any patterns that emerge relating to handling injuries, and take action.
- Make use of support networks in the provision of training, treatment or therapy.
- Ensure all employees are aware of the responsibilities in the workplace regarding accident prevention – and ensure they know what to do in the unlikely event of an accident.
- Support local and national campaigns that raise awareness in promoting a healthy back.

SUGGESTED FURTHER READING

Beiring-Sorensen, F. (1985) Risk factors of back trouble in individual occupations in Denmark, *Ergonomics*, **28**, 51–4.

Health and Safety Executive (1992) *Management of Health and Safety at Work Regulations.*

Health and Safety Executive (1992) *Manual Handling Operations Regulations.*

Pheasant, S. and Stubbs, D.A. (1991) *Lifting and Handling: An Ergonomic Approach*, National Backpain Association, London.

Seccombe, I. and Ball, J. (1992) *Back Injured Nurses: A Profile*, RCN, Discussion Papers.

REFERENCES

Beiring-Sorensen, F. (1982) Low back trouble in a population of 30-, 40-, 50-, and 60-year-old women and men, *Danish Medical Bulletin*, **29**, 289–98.

Beiring-Sorensen, F. (1983) A prospective study of low back pain in a general population: 1 Occurrence, recurrence and aetiology, *Scandinavian Journal of Rehabilitation Medicine*, **15**, 71–9.

Beiring-Sorensen, F. (1985) Risk factors of back trouble in individual occupations in Denmark, *Ergonomics*, **28**, 51–4.

Bell, F. (1984) *Patient Lifting Devices in Hospitals*, Croom Helm, London.

Bell, F., Dalgity, M.E., Fennell, M.J. and Aitken, R.C.B. (1979) Hospital ward patient lifting tasks, *Ergonomics*, **2**, 1257–73.

Brown, J.R. (1972) *Manual Lifting and Related Fields: An Annotated Bibliography*, Labour Safety Council of Ontario, Toronto.

Buckle, P.W. (1983) *A Multidisciplinary Investigation of Factors Associated with Low Backpain*, PhD thesis, Cranfield Institute of Technology.

Buckle, P.W., Stubbs, D.A., Randle, I.P.M. and Nicholson, A.S. (1992) Limitations in the application of materials handling guideline, *Ergonomics*, **35**(9), 955–64.

Clemmer, D.I. and Mohr, D.L. (1991) Low back injuries in heavy industry II: labor market forces. *Spine* **16**(7), 831–4.

Dehlin, O., Hedenrud, B. and Horal, J. (1976) Back symptoms in nursing aides in a geriatric hospital, *Scandinavian Journal of Rehabilitation Medicine*, **27**, 518–24.

Frymoyer, J.W., Pope, M.H., Constanza, M.C., Rosen, J.C., Coggin, J.E. and Wilder, D.G. (1980) Epidemiologic studies of low backpain, *Spine*, **5**, 419–23.

Frymoyer, J.W., Pope, M.H., Clements, J.H., Wilder, D.G., Macpherson, B. and Ashikaga, T. (1983) Risk factors in low backpain, *Journal of Bone and Joint Surgery*, **65**, 213–18.

Garg, A. and Moore, J.S. (1992) Epidemiology of low backpain in industry, *Occupational Medicine*, Oct.–Dec., **7**(4), 593–608.

Garg, A., Owen, B.D. and Carlson, B. (1991) An ergonomic evaluation of nursing assistants' job in a nursing home, *Ergonomics*, **34.**

Greenwood, J.G. (1986) Back injuries can be reduced with worker training reinforcement, *Occupational Health and Safety*, **55**, 26–9.

Harber, P., Billet, E., Gutowski, M., Soohoo, K., Lew, M. and Roman, A. (1985) Occupational low backpain in hospital nurses, *Journal of Occupational Medicine*, **27**, 518–24.

Jensen, R. (1987) Disabling back injuries among nursing personnel: research needs and justifications, *Research In Nursing and Health*, **10**, 29–38.

Jensen, R. (1990) Prevention of back injuries among nursing staff. In W. Charney and J. Schirmer (eds), *Essentials of Modern Hospital Safety*, Lewis Publishers Inc., Chelsea MI.

Kelsey, J.L. (1982) *Epidemiology of Musculoskeletal Disorders*, Oxford University Press, Oxford.

Kelsey, J.L., Githens, P.B., White, A.A., Holford, T.R., Walter, S.D., O'Connor, T., Ostfield, A.M., Weil, W., Southwick, W.O. and Calogero, J.A. (1984) Epidemiological study of lifting and twisting on the job and risk for acute prolapsed lumbar intervertebral disc, *Journal of Orthopaedic Research*, **2**, 61–6.

Klein, B.P., Jensen, R.C. and Sanderson, L.M. (1984) Assessment of workers' compensation claims for back strains/sprains, *Journal of Occupational Medicine*, **26**, 443–8.

Lloyd, P., Tarling, C., Troup, J.D.G. and Wright, B. (1987) *The Handling of Patients: A Guide for Nurses*, 2nd edition, The Royal College of Nursing, London.

Manning, D.P., Mitchell, P.G. and Blanchfield, L.P. (1984) Body movements and events contributing to accidents and non-accidental back injuries, *Spine*, **9**, 734–9.

Margora, A. (1970a) Investigations of the relation between low backpain and occupation 1, *Industrial Medicine*, **39**(11), 31–7.

Margora, A. (1970b) Investigations of the relation between low backpain and occupation 2, *Industrial Medicine*, **39**(12), 28–34.

Margora, A. (1972) Investigations of the relation between low backpain and occupation 3, *Industrial Medicine*, **41**(12), 5–9.

Margora, A. (1973a) Investigations of the relation between low backpain and occupation 4: Physical requirements: bending, rotation, reaching, and sudden maximal effort, *Scandinavian Journal of Rehabilitation Medicine*, **5**, 186–90.

Margora, A. (1973b) Investigations of the relation between low backpain and occupation 5: Psychological aspects, *Scandinavian Journal of Rehabilitation Medicine*, **5**, 191–6.

Nicholson, A.S. and Clark, A.G. (1986) The use of handling guidelines in assessing baggage handling facilities. In W. Karkowski (ed.), *Trends in Ergonomics: Human Factors, III*, Elsevier, Amsterdam, 855–62.

Personick, M.E. (1990) Nursing home aides experience increase in serious injuries, *Monthly Labour Review*, **113**(2), 113–37.

Seccombe, I. and Ball, J. (1992) *Motivation, Morale and Mobility: Qualified Nurses in the 1990s*, IMS Report 233.

Snook, S.H. (1978) The design of manual handling tasks, *Ergonomics*, **21**, 963–85.

Snook, S.H., Campanelli, R.A. and Hart, J.W. (1978) A study of three preventive approaches to low back pain, *Journal of Occupational Medicine*, **20**, 478–81.

Stubbs, D.A., Baty, D., Buckle, P.W., Fernandes, F., Husson, M.P., Rivers, P.M., Worringham, C.J. and Barlow, C. (1986) Back pain in nurses, Ergonomics Research Unit, University of Surrey, Guildford.

Thomas, S.A., Stone, W.E., Greenwood, K. (1990) Australian workers with back injury: a statistical overview, *International Journal of Rehabilitation*, **13**(3), 195–203.

Troup, J.D.G., Martin, J.W. and Lloyd, D.C.E.F. (1981) Backpain in industry: a prospective survey, *Spine*, **6**, 61–9.

Urlings, I.J.M., Nijboer, I.D. and Dul, J. (1990) *Ergonomics*, **33**, 629–37.

Vallfors, B. (1985) Acute, subacute and chronic low backpain: clinical symptoms, absenteeism and work environment, *Scandinavian Journal of Rehabilitation Medicine*, Supplement 11, 1–98.

Van Akkerveeken, P.F. (1985) Teaching aspects, *Ergonomics*, **28**, 371–7.

Videmean, T., Nurmimen, T., Tola, S., Kuorinka, I., Vanharanta, H. and Troup, J.D.G. (1984) Low back pain in nurses and some loading factors at work, *Spine*, **9**, 400–4.

Walker, A.H. (1985) Taking the weight off the delivery man, *Safety Practitioner*, **3**(10), 28–9.

Wickstrom, G. (1978) Effects of work on degenerative back disease, *Scandinavian Journal of Work, Environment and Health*, **4**, supplement 1, 1–12.

USEFUL ADDRESSES

National Backpain Association, 31–33 Park Road, Teddington, Middlesex TW11 0AB.

London Hazards Centre, Headlands House, 308 Grays Inn Road, London WC1X 8DS.

British Chiropractic Association, Registered Office, Premier House, 10 Greycoat Place, London SW1P 1SB.

The General Council and Register of Osteopaths, 56 London Street, Reading, Berkshire RG1 4SQ.

The College of Osteopaths, 1 Furzehill Road, Borehamwood, Hertfordshire WD6.

Coping with substance abuse at work: alcohol and smoking

4

Defining the problem • General effects on the individual's health and social circumstances • Changing roles of women and men in society • Conflicting opinions relating to performances and accidents at work • Some lessons from the USA • Prevention: what role can managers play? • Checklist for managers • Examples of good practice • Passive smoking and its effect on individuals in the work environment • The case for a smoking policy • The physical affects of smoking • Benefits of smoking cessation • Summary • Suggested further reading • References • Useful addresses

DEFINING THE PROBLEM

Alcohol and its effect in the work environment is regarded as a problem if:

- it interferes with the individual's work duties and ability to perform;
- it is affecting other people's work and increases their workload as a result of a colleague being compromised as a result of excessive alcohol consumption;
- the individual concerned is deemed to be a potential liability if they are not kept under control.

Research in the field provides evidence in defining the problem of the misuse of alcohol from many facets. The problem is a broad one as there appears to be some inconsistency in findings. This is especially true in the recording and reporting of accidents where the excess consumption of alcohol while on duty is a contributory factor. There is the additional problem of proving whether or not alcohol is a factor in the reported cases of injuries. What is clear, however, is that it contributes to increased sickness and absenteeism, in the breakdown of working relationships and has an impact on the health of the individual, as well as the reputation of the organization.

SICKNESS AND ABSENTEEISM ISSUES

One in ten of drinking males experience alcohol-related problems at work. This applies at all levels of seniority in organizations. At least 75% of people seeking help and advice for alcohol-related problems are in employment. Up to 25% of accidents at work involve intoxicated workers according to the Brewers' Society. They also indicate that heavy drinkers take up to four times as many days off work as other employees. Drink-related absenteeism is estimated to cost industry at least £800 million a year in England and Wales alone.

> A study of a group of 300 problem drinkers showed that because of their drinking, 66% were frequently late for work, 40% were absent on Monday morning once or twice a month, 21% were absent on Monday morning more than twice a month. It transpired that male drinkers with advanced problems had lost on average 86 working days during the year before going for treatment. In comparison, the average sick leave per person was about 19 days per year.

Other studies have indicated, as reported by the employed persons themselves, a clear trend for tardiness and absenteeism to decline alone with a rise in the relative level of alcohol consumption. This finding has been foreshadowed by earlier suggestions that coming to work regularly and on time is a 'cover-up' strategy utilized by developing problem drinkers who find their drinking highly rewarding and are prone to 'denial' of any aspect of having a drink problem. In other words, work absence and tardiness trigger the need for accounts which may lead to others discovery of problem drinking. Thus it is suggested that presence at work can be an effective screen, even though the employee may then exhibit 'on-the-job absenteeism', meaning disappearances from the work station, lengthy breaks, sleeping or other non-performances that tend to fall outside the official organization's recording of absenteeism and tardiness.

Some commentators in the field of alcohol research state that inappropriate drinking at work can lead to accidents. This in turn disrupts production and can endanger the health and safety of the workforce especially if handling machinery or equipment. Faulty decision-making with impaired concentration can also lead to a slight reduction in the quality of work. Indeed, some British companies estimate that days lost from work costs upwards of 2.5% of its annual payroll.

Glatt and Mills (1965) found that half their clients who were middle aged and older had developed a regular pattern of taking time off work as a result of their drinking, since their early thirties. Later studies found that at least 98% of their clients lost on average 86 days work a year, 66% often being late and 61% frequently absent on Monday mornings. Further findings by Sand and Madden (1976) found that clients still at work were losing an average of 70 days a year; those no longer employed had previously lost no less than 208 days a year.

THE RIGHTS OF EMPLOYEES

There appears to be a lack of clarity in ways of dealing with employees who have a recognized drink problem and have been caught drinking on duty.

To dismiss an employee as soon as a drink problem is brought to an employer's attention will most certainly be unfair. Employers have both common law and statutory duties towards their other employees, members of the general public and other people's employees who are working on their premises. Managers or colleagues who cover up for an employee with an alcohol problem may be contributing to important breaches of their employers' statutory and common law obligations. There is also the question of personal liability under Sections 7, 8, 36 and 37 of the Health and Safety at Work Act 1974. If managers knowingly connive at an offence they could be prosecuted.

Consideration should be given to what the disciplinary rules of a company say about drinking alcohol on duty or being at work under the influence of alcohol. It is accepted by tribunals that employers are entitled to treat such offences as gross misconduct and dismiss anyone who commits such an offence. When employers have lost cases of this nature it has generally been because of some substantial breach of procedure. Employers and trade unions in the type of industry where it would be dangerous for an employee to drink on duty should ensure that the disciplinary rules and procedure spell this out under **gross misconduct** and display notices to this effect.

The manufacturing and chemical industries, construction companies and the Health Service are examples where no drinking rules are extremely important. Drinking offences which lead to breaches of health and safety requirements, and to accidents to other employees or members of the public, could result in negligence claims against the employer.

DISTINCTION IN DEFINITION

Managers need to be aware of the clear distinction between 'drinking offences' and 'alcohol abuse'. It is all too simple to label an individual as an abuser of alcohol at the workplace without proper investigation of the facts and consideration of the capacity in which the individual is employed. Stigma is attached to the term 'abuse' and is difficult to retract once the employee is so labelled. Hence careful consideration is necessary.

Drinking offences should include:

- drinking alcohol on duty;
- possessing alcohol at work;
- being drunk on duty.

Alcohol abuse may be regarded as an illness once the employee is assessed by a doctor and his/her capability to perform duties taken into consideration.

NATURE OF THE PROBLEM

Another part of the equation in defining the problem relates to the availability of alcohol and why there is the need to consume it.

1. Alcohol is available at some workplaces as there is usually a bar in the company's social club. There may also be the 'local' just down the road or around the corner, regularly frequented by employees.
2. There are also social pressures to consume alcohol at work. If colleagues get together at the workplace to discuss business or personal matters, it is often easier to do so in a social context where alcohol is available.

While most people who drink alcohol do so in moderation, there is a significant number whose drinking habits bring harm to their own lives and those around them. It is difficult to estimate the cost in terms of reduced inefficiency and accidents at work, but it is accepted in official circles that it can be costly. Maynard and Godfrey (1994) noted that the estimated social cost of alcohol per year for some countries is as shown in Table 4.1.

Table 4.1 Estimated social cost of alcohol per year

Country	Cost per year
Australia	\$A 6.027 billion
Canada	\$C 5.7 billion
United Kingdom	£2.0 billion
United States of America	\$5.8 billion

(Source: Maynard and Godfrey, 1994.)

As mentioned earlier, the problem from the employer's perspective manifests itself in sickness absenteeism and generalized and frequent absence. Additional difficulties that the employee presents arise with lateness (both morning and after lunch), lost time on the job, inefficiency in alertness, accuracy and judgement.

For the employee, the most advanced drinking problems discovered at the workplace can lead to loss of pay and promotion prospects, injury to himself/herself or others in accidents which may be sporadic or recurrent. Eventually, if the problem remains untreated, the employee could lose his or her job, support of friends and family and self-respect.

Colleagues in the workplace are also affected as they too may be at risk from accidents triggered by the problem drinker. They too will have to cover work poorly completed or neglected. This may become an embarrassing situation if it becomes a regular habit.

In attempting to define the problem, there may also be assessment constraints in:

- lack of awareness regarding the extent of the problem;
- dealing with the problem;
- ability to calculate precise economic loss involved;
- resistance from employees in the guise of 'cover-up', group loyalty, unwillingness to participate in investigations for fear of recrimination;
- inadequate record-keeping;
- inertia: 'the problem will go away'. When the problem becomes too overwhelming, any action then taken may merely be regarded as a reaction to a crisis.

There are ways in which alcohol consumption may threaten the health and safety of the drinker and of others at work. The consumption of even a small amount of alcohol can pose a particularly high risk to safety in certain circumstances, for example, operating a crane or driving heavy goods vehicles. Surveys have revealed that drink problems can occur at any age and within any social group. The reasons are as varied as the people affected but include attempts to:

- cope with stress;
- cope with unbearable situations;
- use as a mask against shyness or weakness.

People may also simply become accustomed to drinking through their work or social connections. There is therefore no single 'alcoholic personality'. However, some groups of people are more at risk than others. For example, young people often associate drinking with toughness and maturity. In practice, this group is more likely to be involved in drunkenness than addiction.

Those working in 'high-risk' occupations also become accustomed to alcohol as a normal part of their duties. They may be in jobs which bring them into contact with situations where the work is highly pressured, where much travel is involved, where heavy drinking is the norm, where the individual is frequently away from home. The occupations that come into these categories include journalism, media work, commercial travelling, sales, the entertainment industry and business.

In conclusion, the problem of excessive consumption of alcohol at work is a multi-faceted one in which managers need to find common ways of dealing with its causes and effects if they are to have a healthy workforce that is operating in unison to improve the culture of the organization.

GENERAL EFFECTS ON THE INDIVIDUAL'S HEALTH AND SOCIAL CIRCUMSTANCES

Alcohol is absorbed in the bloodstream on consumption. Almost all the alcohol is burnt up by the liver and the remainder disposed of either by sweating or in the urine. The concentration of alcohol in the body depends on how much is drunk, whether the stomach is empty or not, height, weight, age and gender. If a person is smaller or lighter in weight than average, or not used to drinking alcohol, they are more likely to be affected by alcohol.

Alcohol is a **depressant**. It acts by depressing certain brain functions, therefore affecting judgement, self-control and co-ordination. Other effects on the body include the following.

DEHYDRATION

Alcohol tends to make water move out of the body's cells which leads to dehydration and other factors producing a hangover. It is not possible to sober up by drinking black coffee or a cold shower as it takes up to one hour for each unit of alcohol consumed, i.e. half a pint of beer, to be processed by the body.

OVERWEIGHT

Alcohol contains many calories. Heavy drinkers can be overweight and still suffer malnutrition. Energy is obtained from alcohol instead of food, which lacks essential vitamins and nutrients.

DISEASE AND DISORDER

Excessive alcohol consumption can lead to stomach disorders, high blood pressure, vitamin deficiency, sexual difficulties and nervous disorders. In more extreme cases, excessive drinking can lead to cirrhosis of the liver, hepatitis, cancer of the mouth, throat and oesophagus.

INTERFERENCE IN MENSTRUATION

Alcohol affects women more quickly than it does men and the effects usually last longer. This is because the concentrations of alcohol vary and are retained for longer periods in the bloodstream at different parts of the menstrual cycle. Concentrations are particularly greater just before ovulation and a few days prior to menstruation.

The effects of alcohol on the body may be short or long term depending on the frequency, quantity and regularity with which alcohol is consumed.

In the short term, alcohol produces alterations in mood. Its effect on the nervous system is to reduce control of the brain and muscles. Judgement suffers and reflexes are slowed down. Before an individual

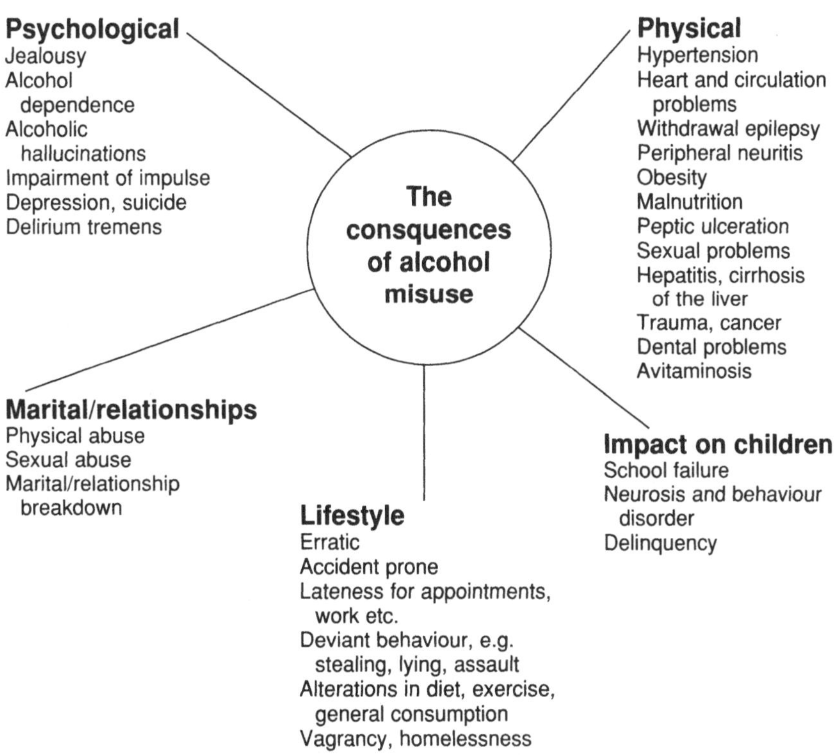

Figure 4.1 The consequences of alcohol misuse

realizes that they have had too much alcohol to drink, they often feel fine. This is considered the dangerous phase, because when judgement, muscular control and reflexes are impaired, accidents can and do occur at work, home and on the roads.

Long-term effects of excessive alcohol consumption are less obvious and regularly underestimated. The effects usually result from the strain alcohol places on the body when it is forced to cope with regular heavy drinking. Pancreatitis, hepatitis, gastritis, and various cancers can develop over a period.

Further effects are identified in Figure 4.1.

CHANGING ROLES OF WOMEN AND MEN IN SOCIETY

More women are consuming increased quantities of alcohol. In the USA alone, of the estimated 15 million alcohol-abusing or alcohol-dependent individuals, at least a third are women (NIAAA, 1990). Public and scientific concerns over women's drinking problems have grown over the last 20 years. This concern is fuelled by changes in sex-role attitudes and in the work and family roles of women which suggest that male and female drinking patterns may become similar (Roman, 1988c).

Role changes that may expose women to more drinking occasions include increased labour-force participation, an increase in the proportion of single or divorced adult women, and later childbearing. Beiner (1987) points out that concerns relating to a projected convergence in male and female drinking behaviour are compounded by the belief that when women and men exhibit similar drinking behaviours it will produce more difficulties for women. This belief is grounded in:

1. biological differences between men and women;
2. societal norms against female intoxication; and
3. the integration/ambivalence notion that those women who ignore the social norms against female intoxication are especially vulnerable to loss of control over their drinking.

Heavy drinking is usually associated within certain social circles with risk-taking, stamina and seeing through the social pretence as well as defying social convention. These qualities are more admired in men than they are in women.

Despite the changing roles of women in society, particularly in the choices of work, research shows that it is nonetheless males who are more likely to display or become involved in problems relating to alcohol. Indeed, Robbins (1989) found that alcohol intoxication was more firmly associated with males than females. Males exhibit higher incidences in unsafe driving, trouble with the police and problems at work than women. In addition to these findings, Robbins also points to the fact that alcoholism is more frequently associated with job loss, accidents and violence. It is also the case that men are more likely to be referred for treatment following legal, financial or job difficulties, while women are more likely to seek assistance voluntarily (Gomberg, 1979; Knupfer 1982).

CONFLICTING OPINIONS RELATING TO PERFORMANCE AND ACCIDENTS AT WORK

Studies carried out in France on employees in a range of industries found that those in serious accidents, losing 15 days or more from work, had a history of higher alcohol consumption than people having only minor accidents. The studies also highlighted that employees with blood alcohol levels of 30 to 100 mg per 100 ml experienced more accidents than people with less than 30 mg per 100 ml. This has been supported by two UK studies where alcohol-related accidents tended to occur after lunch and between 22.00 hours and 24.00 hours, and finally between 08.00 hours and 10.00 hours. These were associated with shift patterns, pub opening times and finally hangovers, with respondents retaining a high level of alcohol in their bloodstream while on duty (Argyropoulos-Grisanos and Hawkins, 1983; Beaumont and Allsop, 1983).

Webb *et al.* (1994) showed there was a significant relationship between injury-related absences and problem drinking; there was no such relationship between absences and high levels of alcoholic consumption. However, so-called 'problem drinkers', i.e. those found to experience low job satisfaction and older workers, were found to have an increased incidence of injury-related absences than other workers.

The above studies conflict with the majority of occupational health and safety statistics which provide information relating to blood alcohol concentrations (BACs) of fatally injured workers alone. Indeed, between 4.3% and 11% of fatally injured workers in the coroner's data studies had BACs of at least 80 mg per 100 ml, while between 0.6% and 9.2% had BACs of at least 100 mg/100 ml. There are studies also which have found no relationship between work injuries and high alcohol consumption (Powell *et al.*, 1971; Hertz and Emmett, 1986; Beaumont and Hyman, 1987; Buchanan, 1988).

The field of alcohol studies in the workplace is still to be developed. Indeed, there is scope for new work employing the following.

- Screening of an entire workforce for problem drinking and its contentious implications, using adequate screening measures.
- Use of adequate measures and recording of all industrial injuries regardless of severity.
- Use of a control or comparison group.

- Selection of adequate sample size and prospective collection of data on accidental injuries.

Maynard and McDonnell (1985) have also pointed to the necessity for such a study in order to fill significant gaps in the epidemiological literature related to the costs of alcohol abuse. Until there is a consistent approach to ascertaining information relating to accidents in the workplace directly relating to alcohol consumption, then the problem will continue to be dealt with in an *ad hoc* manner.

SOME LESSONS FROM THE USA

The basis of Employer Assistance Programs (EAPs) stem from the 1940s where substandard and unsatisfactory job performance highlighted the need for intervention to assist problem drinkers. In their early days, EAPs had a very poor image of serving blue-collar workers with alcoholic problems. They suffered from the stigma of associating itself with social disadvantage.

It was not until the 1970s, following restructuring, that EAPs gained popularity. This was allied to a change in US norms influencing organizational management. There was widespread involvement of employers in strategies designed to alter health-related behaviours of employees at all levels. Allied to this was the recognition of increased healthcare costs borne by employers both through direct payments and through payments for employee's third-party coverage for healthcare costs. Health promotion too, saw its advent in the workplace where employers became aware of the importance of health and fitness of employees.

EAPs were started with government assistance and it was the establishment of the National Institute on Alcohol Abuse and Alcoholism (NIAAA) which led to the adoption of a principal means of 'mainstreaming' management and treatment of alcohol problems into the US healthcare system. The NIAAA's interest in the development of a comprehensive EAP model led to the provision of funding in the form of grants. This led to the establishment of a foundation for a set of specialist occupations centred around EAP diffusion and implementation.

Sonnenstuhl and Trice (1986) define EAPs as job-based programmes operating within a workplace for purposes of identifying 'troubled employees', motivating them to resolve their troubles, and providing a means to counselling or treatment for those employees who need these 'services'. In fact, Scanlon (1986) notes that EAPs may be called 'wellness programmes' in that the emphasis is on health rather than on problems.

The purpose of EAPs is an attempt to provide early intervention for behavioural problems before they become costly for the employer and the employee. They operate by channelling the employee to a source of assistance selected with the guidance of an EAP counsellor which will provide the most cost-effective help for the employee's problem.

EAPs provide an opportunity for employees to recover from the difficulties of substance abuse. As well as catering for employees with drink problems, the service extends to drug addiction, family discord and psychiatric illness. The employee's job and career are preserved, and he or she can resume responsibilities once an effective level of accommodation is reached.

Support for the programme has emerged in many occupations and from trade unions. Airline pilots, construction workers and longshoremen have established their local internal EAPs. They serve to meet employee health and welfare goals and have support of unions and management, as EAPs are enshrined in company policy. Employers have to be convinced at the outset that EAPs are going to be beneficial to their organizations. Substantial organizational knowledge and appropriate legal guidance are given before collective decisions are made.

The highest levels of activity of EAPs are in Minnesota, New York, Michigan, California and North Carolina. Recent estimates highlight that as many as 40% of the American workforce have access to EAPs. In recent years, there have been high trends in the rates of self-referral within EAPs. Employees are guaranteed confidentiality and the EAP co-ordinator cannot contact the supervisor for information about job performance unless the employee grants permission. Within this context, the EAP co-ordinator is limited to utilizing the employee as the source of information concerning job performance. One problem that can arise, however, is that self-reports of job performance problems are much less likely to indicate serious performance difficulties than if this information was sought from the employee's supervisor or colleagues (Blum *et al.*, 1993). It is more favourable if the employee's supervisor becomes involved at an early stage as evidence relating to job performance problems may be important in motivating employees to seek treatment or otherwise alter their personal drinking behaviours.

The programmes are varied depending on the needs of employees and the organization; however, for the most part they have similar characteristics:

- availability to employees and their dependents;
- unlimited usage;
- no cost to the user;
- available at times convenient to the user;
- answering service;

- confidentiality;
- professional service provided by professionals;
- follow-up service available;
- outreach service available;
- supervisory training provided.

The programmes can occur on site, in conjunction with the occupational health service, or alternatively, off site for the small or medium-sized company where more privacy is likely to be offered.

EAPs are developing slowly in the UK, their main emergence being in the late 1980s. The effects of stress on performance, and a developing climate of disillusion with aspects of the NHS, appear to be reasons for company conversions (Lee and Gray, 1994).

EAP currently in operation in the UK was developed jointly by Care Assist, a subsidiary company to Royal Insurance and Psychology at Work (an initiative of the Department of Psychology at the Institute of Psychiatry, London University, attached to Maudsley Hospital). The programme offers counselling across a full range of stress-related problems including alcohol, marital and substance abuse issues.

The service is professionally supported by lawyers, counsellors and psychologists who assist employees and their families when required. Back-up services are also provided for managers, and those involved in occupational medicine by way of briefing sessions and training to deal with problems that may arise, as well as informing of policy changes.

PREVENTION: WHAT ROLE CAN MANAGERS PLAY?

Managers should carefully consider the implications and potential difficulties for the employee with a drink problem before introducing measures to curb the difficulties regardless of their status. Employees in all cases should be encouraged to seek assistance and treatment where necessary. Every case is to be considered in relation to circumstances for, as noted earlier, there is no single causative factor in the development of drink problems.

It is imperative that employees have guidance and that there is clarity of policy throughout the organization. Indeed, if possible, the rules concerning alcohol in the organization should be agreed between trade unions and employees and embodied in a written agreement. Managers should also allow a degree of flexibility in the application of the rules along with a clear line regarding penalties, i.e. employees should be in no doubt whether or not they will face dismissal for an initial offence.

> Mr. Beverstock was a delivery driver who returned to his workplace for his overalls at 7.10 pm smelling of alcohol. His managers dismissed him the next day when it was discovered that he had been drinking in a pub at 4.00 pm when he should have been carrying out his delivery duties. The company argued that his offence was worsened by the fact that as he worked alone his conduct could not be monitored and he was therefore in a position of trust and that whilst others were found drinking on site or arriving smelling of alcohol, the offence amounted to a breach of trust.
>
> The dismissal was unfair because the rules of the organization surrounding alcoholic offences were unclear and were not included in employee's contract.
>
> (*Dairy Produce Packers Ltd* v. *Beverstock 1981*, IRLR)

To underline this point, the Department of Employment (UK) notes that:

> a reasonable employer may impose differing standards with regard to the question of alcohol, dependent upon a number of factors such as the nature of his business, the effect which this conduct e.g. drinking, may have upon public opinion, etc. Where it is considered necessary to have specific penalties attached to the misuse of alcohol in a particular enterprise, or where certain types of offences are to be treated in an exceptional manner, this should be clearly laid down and made a term of conduct.

Managers must also be wary of their position. They can also be subject to disciplinary action or dismissal. Where no drinking rules exist, managers are expected to apply them. If they ignore rules by allowing employees to drink on duty, managers can also expect to be dismissed. See box on top of page 72.

Where organizations identify that disciplinary rules need to be tightened, for instance, making the consumption of alcohol a serious misconduct offence, it must be ensured that every employee receives a copy of the new rule and that it is explained to all employees. Employees should be given at least four weeks' notice that such a rule will come into force. Where trade unions are recognized, the new rule should be agreed with them.

When an employee is considered to have infringed the disciplinary codes of the organization because of drink, it is necessary to allow an opportunity to respond fully and thoughtfully to any allegations. This

Mr. Williams, a plant manager at the Ciba-Geigy company overlooked any possible disadvantage to a Christmas drinking party at one of their chemical plants. Employees had all received a letter outlining that there was to be no drinking on the site on Christmas Eve. It also indicated that anyone caught drinking on the premises would be dismissed.

After the festive season had ended, cans and bottles were found on site but by that time it was too late to identify those involved. Mr. Williams was then suspended and then dismissed.

The company's decision was upheld at tribunal to dismiss on the grounds that senior management was expected to act competently and investigate breaches of the employer's policy.

(*Williams* v. *Ciba-Giegy 1981*)

An ovensman working the night shift at a biscuit factory was dismissed after he was seen to be drunk whilst on duty. The company had very strict rules which were known to all employees concerning the consumption of alcohol which had been imposed for safety reasons, and [were] agreed by the union.

(Alcohol Concern, 1991)

usually includes permission for the employee to be sent home on paid leave, and delaying conducting any formal proceedings until the employee is sober.

Once it is recognized that the employee has a drink problem, the condition should be dealt with in a similar manner to an employee who is sick. There should be encouragement for the employee to seek the opinion of the company doctor with the employee's written consent if possible.

The employee should be given opportunities to discuss the problem (and offered counselling), and if necessary, time off with pay. Managers should be aware that an investigation of employees circumstances is likely, because it may be a symptom of some serious medical problem. The employer's commitment is not expected to be finite when the employee experiences a relapse. The industrial tribunals have made it clear that where an employee has a drinking problem which extends over a considerable period of time, it may not be considered unfair to dismiss that employee if s/he continues in the same vane.

> A Leeds (UK) industrial tribunal held that it was not seen as unfair for an employer to dismiss an alcoholic employee who had received considerable assistance from the organization over a long period of time. The individual concerned was regarded as an experienced and capable mechanic who suffered from a nervous condition which led to his uncontrollable drinking behaviour. His managers had been particularly supportive in allowing sick leave and encouraging him to seek medical assistance; financial help was also granted. However, his performance declined and he was seen to be continuing to drink in his car after his work duties were complete.
>
> (*Barber* v. *Tricentol Cars (UK)*)

There is a major difficulty in getting managers to see alcohol as an issue affecting worker's performance. Managers concerned to develop a health-promoting workplace will look at the environment and at the policies and systems within the organization. Managers should also look at procedures and relationships to assess their impact upon the health and therefore the productivity and safety of the workforce.

Surveys indicate that successful programmes may produce benefits such as reductions in accidental injuries and absences, with resultant cost savings. Also, high injury rates among problem drinkers suggest a need to identify employees who may be at risk in order to take preventive action. The workplace is a suitable venue for intervention into alcohol-related problems, because of the likelihood that the employee will be amenable to treatment if his or her job security is at stake. In addition, rehabilitation of the problem drinker may be less expensive than alternatives such as dismissal, early retirement or continued employment of an employee who is not fully productive.

Alcohol prevention programmes themselves need to be specifically devised for the problem drinker rather than merely a general programme for those consuming large quantities of alcohol without an identified uncontrolled behaviour. In a recent study, the best predictors of problem drinking found were high numbers of recent stressful life events, single, separated or divorced marital status, and neuroticism. These findings may assist in the identification of problem drinkers and the development of appropriate treatment programmes (Webb *et al.*, 1994).

CHECKLIST FOR MANAGERS

- **Identify the problem** – investigate complaints and record incidences. Keep records and statistics of accidents and undertake analysis. Break them down into those which are possibly alcohol related and those which are not. Note the times of day when they occur.
- **Make a point of investigating all injuries and accidents** – ask whether the individual has been drinking. Consider the possibility of offering a blood test to any employee who appears to be under the influence of alcohol; when questioning, ask about last night's drinking.
- **Establish links with key individuals within the organization** – i.e. personnel officer, union representatives, occupational health nurse and quality manager. Make the alcohol issue a part of all supervisory and management roles.
- **Consult, advise, involve unions in policy decision-making from the outset rather than presenting a ready-made policy that may lead to difficulties.**
- **Set up a joint committee to conduct campaigns, produce publicity, provide information, monitor and evaluate projects.**
- **Relate the policy to alcohol** – define boundaries. State whether or not alcohol is allowed to be consumed on premises, during working hours and limitations. Devise a means of dealing with offenders; means coping with individual difficulties and the provision of support systems.
- **Balance any perceived loss of personal autonomy with a caring, positive programme of support for the problem drinker.**
- **Emphasize that the policy is in the interest of safe, efficient working and the health of everyone concerned.**
- **Ensure there is a whole organizational approach operating fairly and consistently from top to bottom.**
- **Managers should set a good example by offering non-alcoholic drinks to visitors and at board meetings.**
- **Define 'fitness to work' and put in a policy statement for all to see.**
- **Managerial commitment to organizational policy is crucial.**
- **Provide training for supervisors and middle managers dealing with employees with suspected alcohol problems.**
- **Provide information and access to services.**

EXAMPLES OF GOOD PRACTICE

ROLLS ROYCE UK

This company was one of the first in Scotland to establish a caring policy for employees known to be chronically dependent on alcohol during the late 1970s. Their policy at the time was unwritten and informal and operated mainly through a network of identification and support managed by the unions. Shop stewards argued that they were usually the first to notice signs of dependency in fellow workers and that they were best able to confront them with it. The company provided transport and time off for a union official to take the employee to an external counselling agency.

This policy has now developed into a more formal written agreement, not broadly advertised, and its implementation has been low-key. However, the policy is designed to cope with the person with a problem.

BRITISH STEEL

This company is known for its tight control of alcohol consumption among employees. Employees are sent home if considered unfit for work. The occupational health department tends to identify people with repeated injuries, in an attempt to spot the 'problem drinkers', rather than looking at all accidents to note whether any of them might be alcohol related.

Most accidents occur in the hours of darkness and are dealt with by the occupational health department. For identification of workers 'under the influence' of alcohol, they have to rely on being able to notice unusual behaviour, even though they recognize that this could not be noticeable until the level of intoxication is quite high. Control of alcohol on the premises is noted as excellent in spite of there being a number of contractors and the impossibility of searching people entering the site.

They no longer have problems with drinking in the workplace. There are no longer cans and bottles lying around in the factory, as there sometimes were in the past. Over the last few years, the 15–20 employees who have accepted help under the company's policy have recovered sufficiently to return to work and are still in active employment.

PASSIVE SMOKING AND ITS EFFECT ON INDIVIDUALS IN THE WORK ENVIRONMENT

Tobacco smoke is known to lead to contamination of the environment. Most non-smokers have some quantity of smoke product in their bodies because of exposure to other people's smoke. The exposure for many begins in the home during childhood and continues into adult life in the workplace and during social and leisure activities.

Evidence exists of acute respiratory effects on people's health who are exposed to passive smoking. Indeed, the effects are marked for those particularly susceptible, e.g. asthmatics. Managers who make little or no provision for adequate ventilation and control of environmental pollution as a result of cigarette smoke are inadvertently contributing to this process.

Passive smoking may produce initial symptoms such as headaches, sore throats, sore eyes and loss of concentration from short and prolonged exposure. These symptoms are likely to occur in confined work areas or in large areas such as open plan offices in which many people work together. It is increasingly likely to be these acute symptoms, rather than the more long-term risk of cancer, which are the cause of employees' complaints and necessitates the need for employers to take action.

Indeed, in the late 1980s, a former employee of Ansett Airlines, Australia was awarded $A 20 000 in an out-of-court settlement approved in the New South Wales Compensation Court. The employee's claim was upheld on discovering that she was persistently exposed to tobacco smoke in a confined telephone switchroom over a period of six years which led to the development of symptoms including headaches, nausea and respiratory irritation. The employee's general practitioner confirmed that her health was being compromised as a result of this exposure.

Passive smoking can trigger symptoms for people with angina and asthma. Regular exposure to tobacco smoke has been linked with impaired lung function, the development of lung cancer, heart disease and cancer. Passive smoking at work, when pregnant increases the risk of low birth weight which relates to increased infant mortality.

Smoking-related diseases are also prone to exacerbation in the workplace especially when in contact with:

- dust and particles, e.g. asbestos, cement, silica grain and textiles;
- radioactive substances such as uranium or radioactive gases encountered during coal mining;
- chemical agents: these may include toxic metals such as arsenic and lead; also irritant gases such as sulphur dioxide and chlorine or organic gases such as hydrocarbons.

There are also additional risks if cigarettes become contaminated with the substance leading to a chemical reaction producing a dangerous agent. For example, the breakdown of the product Polytetra fluoroethylene (PTFE) may cause 'polymer fume fever' among smokers. It is reasonable to assume that passive smoking would also increase the effect of other occupational hazards, but so far there has been little research to prove this.

THE CASE FOR A SMOKING POLICY

There has been increasing cause for managers to take action to restrict smoking in the workplace:

1. to prevent liability for causing injurious health due to exposure of employees unnecessarily to the risk of tobacco smoke;
2. to take advantage of better deals with insurance companies;
3. to have a more healthy workforce, thus reducing sickness and absenteeism.

> In 1985, the Swedish Court of Appeal ruled that a case of lung cancer in a non-smoker should be classified as an occupational injury due to passive smoking in the workplace. The case involved a lifelong non-smoker who had been exposed to tobacco smoke in a poorly ventilated office for eighteen years. She died at the age of 54, two years after a diagnosis of lung cancer. The Appeal Court upheld a decision made by the Regional Insurance Court that compensation should be paid, based on an assessment of the case in terms of exposure to workplace agents.

This definitive case made it absolutely clear worldwide, that action was required to prevent further exposure to tobacco smoke in the workplace. Indeed, it led to further reporting of similar cases.

1. An employee of Bristol City Council, UK, had been experiencing passive smoking through sharing an open plan office with several heavy smokers. She suffered from a chest condition which became worse due to tobacco smoke. In 1986, she filed a formal grievance against the Council for the right to work in a smoke-free environment. The Council did make some attempt to reduce her exposure to the smoke by providing fans, opening windows and other similar measures which had all proved inadequate. Although the employee was aware that the Council had entered into negotiation with the union representatives of the organization concerning the development of a smoking policy, she filed a grievance. The grievance was based on her individual suffering while waiting for a Council-wide policy to be agreed. She was supported by her union and the grievance was upheld. As a result, the Council was obliged to provide her with a private office.

2. In Australia, cases have been brought by non-smokers against their employers. In each case, the employer has had to pay compensation, find a work environment free of tobacco smoke or both. To avoid leaving itself open to further cases, the Australian Public Service Board announced a policy of achieving smoke-free workplaces in all government offices. This decision followed a ruling by the Attorney-General that department secretaries had the legal power to ban smoking and a finding by the Medical Review Panel of the Department of Defence that the Australian government will find itself involved in escalating numbers of claims unless a firm policy is instituted in Commonwealth buildings to prevent tobacco smoking.

3. In 1986, a West Virginia State Civil Service grievance board ruled in favour of a non-smoking employee, stating that it was clear that existing ventilation, smokeless ash trays and filters did not adequately protect non-smokers. The board ruled that the only way to protect employees was to prohibit all smoking in the building.

Consideration had been given to the matter of passive smoking and the harm it causes to employees previously in some workplaces. Some action has been employee led. In a major US study of 10 000 non-smoking office workers, it was found that 50% had difficulty working near smokers, and 36% said they regularly had to move from their work station to escape the smoke. The same study found that 25% of non-smokers very frequently or always reacted to tobacco smoke with frustration. A similar proportion felt hostile to smokers or management. Approximately one in seven of those who had difficulty working with smokers said they had used sick leave during the previous twelve months because of the tobacco smoke around them at work.

Another US survey found that over a third of non-smokers had left their work areas at some time in order to escape cigarette smoke. About a quarter felt frustrated about the problem or hostile towards smokers and management.

In the UK, the Health and Safety Executive (HSE) has been concerned about passive smoking for some time. It has adopted a policy of encouraging employers to limit the exposure of non-smokers to tobacco smoke at work on a voluntary basis. This is carried out either through workplace agreements which restrict smoking to certain areas or provide for segregation of smokers and non-smokers or by providing ventilation. The HSE has been unwilling to commit itself concerning the use of the Health and Safety at Work Act to control smoking at work, although as long ago as 1979 the HSE expressed the view that: 'The fact that smoking is a personal habit does not of itself exclude the matter from the scope of the Health and Safety at Work Act Section 2. Passive smoking may be relevant to the welfare and subject to medical evidence, the health of the people at work.'

Currently, the UK position does not stipulate any legal requirements for employers to have a no-smoking policy; however, more general legislation does relate to the issue as follows.

1. Employers have a **duty of care** under Common Law.
2. The Public Health Act 1936, requires employers to **provide sufficient ventilation** and keep the air free of **noxious effluvia** which may be harmful to health.
3. Section 2 (2) of the Health and Safety at Work Act 1974 requires an employer to provide for employees an environment which is 'so far as is reasonably practicable, safe . . . And adequate as regards facilities and arrangements for their welfare at work'. Also Section 2 (6) of the Act requires employers to consult with safety representatives on developing and monitoring health and safety measures in the workplace. The Health and Safety Executive has assisted in this process.
4. The Control of Substances Hazardous to Health Regulation 1988 (COSHH), has strengthened the 1974 Act in identifying the fact that if tobacco smoke is thought to present a significant hazard, an assessment is required and the principles of COSHH make elimination rather than control of the hazard a primary solution.

In the United States, courts have ruled that employers have a common-law duty to protect non-smokers from the hazard of second-hand smoke. In addition, the Court of Appeal in California has ruled that employees cannot be discharged or discriminated against because they complain in good faith about workplace conditions that are regarded as unsafe. The courts have also supported claims by non-smoking workers for disability benefits or retirement if workers are extremely sensitive to cigarette smoke and a smoke-free environment is not available or not offered.

Insurance companies in the US have also enhanced this legislation. The National Association of Insurance Commissioners adopted a resolution calling for different health insurance rates for smokers and non-smokers. As a result, some companies extend lower health insurance rates to businesses with non-smoking policies or to small firms where no employees smoke.

The decision to control smoking at the workplace has been difficult from a legislative perspective. However, legislation to reduce the risk of disease from smoking and passive smoking now extends throughout 13 countries of the WHO European Region. These countries include Austria, Belgium, Bulgaria, Czechoslovakia, France, Germany, Iceland, Norway, Poland, Romania, Spain, Sweden and Russia. Bulgaria already prohibits smoking at workplaces where any non-smoker is employed and without exception where expectant or nursing mothers are employed.

In May 1986 a resolution was passed by the Thirty-Ninth Assembly of the WHO in Geneva urging that Member States should implement smoking control strategies to:

> ensure that non-smokers receive effective protection, to which they are entitled, from involuntary exposure to tobacco smoking in enclosed public places, restaurants, transports and places of work entertainment.

The resolution states that inhaling other people's smoke 'violates the right to health of non-smokers who must be reported against this noxious form of environmental pollution'.

Support for this resolution came in March 1988 when the British Government's Independent Scientific Committee on Smoking and Health published its Fourth Report (The Froggatt Report) which concluded that

> people who have never smoked but who have been exposed to environmental tobacco smoke through most of their lives have a 10% to 30% higher risk of lung cancer than non-smokers not so exposed . . . [this] might amount to several hundred out of the current annual total of about 40,000 lung cancer deaths in the UK . . .

Clearly then, steps to reduce the hazards of smoking at work should be on any manager's agenda. Regardless of the extent to which the law presently covers the problem of smoking at work, it is simply not wise for any employer to wait until a court decides there is a hazard. As with other cases of protection from workplace hazards, it is best for employers to be in advance of legal developments. If an employer is challenged, the tribunal will be looking at what a reasonable employer' would do.

Some companies have been far-sighted enough to have informal policies on smoking, claiming that they rely on the courtesy and consideration of smokers toward non-smokers. However, where there is no formal policy, non-smoking employees may be unwilling to assert themselves for fear of being branded trouble-makers. New staff can also present a challenge to traditional arrangements. A formal policy is the best way forward as it frees managers from *ad hoc* arrangements where smokers and non-smokers have difficulty working together. If employees are presented with clarity of formal arrangements then conflict is reduced.

THE PHYSICAL EFFECTS OF SMOKING

Tobacco smoke contains over a thousand chemicals which are released into the air as gas and particles. There are two main groups of tobacco smoke: **mainstream** and **sidestream** smoke.

MAINSTREAM SMOKE

This is smoke which is inhaled directly by the smoker and exhaled into the environment.

SIDESTREAM SMOKE

This is released in larger quantities from the smouldering tip of the cigarette between puffs. 85% of cigarette smoke is sidestream which contains harmful substances including tar, polycyclic monocarbons, carbon dioxide, ammonia and nitrosamines. The smoker inhales both mainstream and sidestream, and the non-smoker sidestream and exhaled mainstream smoke of the smoker.

Sidestream smoke is known to be as toxic as mainstream smoke, although tar from sidestream smoke has been shown to be more carcinogenic per unit weight than mainstream tar. Examples of substances in smoke include the following.

- **Nicotine:** the drug which stimulates the nervous system. It increases the blood pressure making the heart work harder. It is an addictive substance known to raise the levels of fatty acids in the blood, and thus increases the likelihood of clot formation and restriction of blood flow to tissues.
- **Carbon monoxide:** a poisonous gas which combines with haemoglobin in the red blood cells, thus inhibiting the flow of oxygen to the tissues of the body.
- **Tar** forms a sticky brown substance in the lung and air passages. It contains many chemicals some of which are carcinogenic. Tar, coupled with soot and ammonia, accumulates in the lungs and windpipe leading to bronchitis and emphysema. Heat from the burning cigarette can add to lung damage and windpipe damage by burning away the protective lining and cilia of the windpipe, normally responsible for trapping dust and bacteria in inhaled air. As a result of cigarette smoking, dust, dirt and mucous collects in the air passages initially leading to smoker's cough. Prolonged smoking results in damage to the air sacs resulting in emphysema.

In the UK, smoking is responsible for a third of all cancer deaths, 88% of chronic obstructive lung disease deaths and 13% of cardiovascular deaths. Smokers have twice the number of deaths due to coronary heart disease and 10 times the risk of developing lung cancer. Smoking is responsible for 90% of all lung cancer and more than 80% of emphysema and chronic bronchitis. The latter two conditions are major causes of absenteeism.

Heart disease kills more smokers than lung cancer. About one in four coronary heart disease deaths are due to smoking. Below the age of 45, men who are heavy smokers have 10 to 15 times the risk of fatal heart attacks than non-smokers. It has been estimated that it can cost an employer £250 000 to replace a 45-year-old executive lost through a fatal heart attack. The cost to companies which invest heavily in staff training can be even greater.

Smokers may be less productive because of reduced concentration, eye irritation and the time taken up by the ritual of smoking. Estimates of time lost to the smoking ritual range from about 8 minutes per day per smoker to 15 to 30 minutes per hour. The nature of the work affects how much of this lost time could be argued to be 'creative thinking', but for some jobs, work simply stops the employee from lighting up. Variations in performance may also occur as a result of smokers' experiencing withdrawal symptoms between cigarettes. Several businesses have lost 2–10% of their efficiency; indeed, there is on average a loss of 30 minutes work per day, that is, 6% of the work done for a year. Smokers also take one-third more sick leave than non-smokers.

BENEFITS OF SMOKING CESSATION

The introduction of a no-smoking policy in the workplace can have the following benefits.

IMPROVED GENERAL WORKPLACE HEALTH

A survey of Pacific Northwest Bell Telecommunication in the USA found that about a third of the smoking employees attended cessation programmes, about a third were still not smoking between 1983 and 1989, and about four-fifths felt the policy had assisted them in stopping.

Another US survey found that after a policy's introduction, smokers employed in a general hospital were using on average four cigarettes a day less. If applied to all British workplaces, this would be equivalent to a 9% reduction in total cigarette consumption with obvious benefits in reducing smoking-related disease.

IMPROVED SAFETY

An estimated 6.5% of all fires on individual premises arise from smoking materials.

LOW MAINTENANCE COSTS

Smoking increases litter and necessitates more frequent renewal of decoration and furnishings. Savings through increased ventilation, cleaning and maintenance costs are also made. A hotel chain in the US reported that after becoming a no-smoking establishment, rooms could be cleaned 26% faster and spring-cleaned 41% faster and upholstery cleaning costs were reduced by 90% (Sanders, 1990).

REDUCED ABSENTEEISM DUE TO SICKNESS

As indicated earlier, smokers take more time off work due to sickness than non-smokers.

INCREASED EFFICIENCY AND PRODUCTIVITY

Atari Computers, Germany, rewards non-smokers an additional six days annual holiday in recognition of additional productivity. Perhaps managers could follow their example or offer other incentives.

LOWER RISK OF LOSING KEY EMPLOYEES THROUGH PREMATURE RETIREMENT OR DEATH

On average, smokers are twice as likely to die before the age of 65 as non-smokers (provided the non-smoker is not inadvertently subjected to cigarette smoke). This has an effect on labour turnover, recruitment and training costs.

SUMMARY

Alcohol in the workplace is problematic as it interferes with employee performance. If the problem becomes exacerbated, it can have a marked effect on productivity, slows other employees down because they have to deal with the additional work their colleague has failed to complete and it will undoubtedly affect the health of the employee. It will almost certainly lead to conflict if not dealt with.

There are a range of issues that are encapsulated in defining the problem and how to deal with it which includes sickness/absenteeism, rights of employees, effects relating to personal circumstances and the changing roles of men and women in society. Assistance for employees and policy issues to protect employer and employee are regarded as allies in prevention and methods for coping with such issues as they arise.

Passive smoking in recent years has been a major issue for managers at work as there has been an increase in claims by non-smokers whose health has been affected as a result of exposure to unwanted cigarette smoke.

Progressive employers are taking policy issues on board and adjusting the workplace to accommodate their workforce, thus fostering a healthier, cleaner environment.

SUGGESTED FURTHER READING

Alleyne, B.C., Stuart, P. and Copes, R. (1991) Alcohol and other drug use in occupational fatalities, *Journal of Occupational Medicine*, **33**, 496–500.

Blum, T.C., Roman, P.M. and Martin, J.K. (1993) Alcohol consumption and work performance, *Journal of Alcohol Studies*, **54**, 61–70.

Imperial Cancer Research Fund/Cancer Research Campaign (1991) *Passive Smoking – A Health Hazard*.

Institute of Personnel Management (1990) *Smoking Policies at Work – An IPM Guide*, IPM, UK.

REFERENCES

Alcohol Concern (1991) *Workplace Advisory Service*.

Argyropoulos-Grisanos, M.A. and Hawkins, P.J.L. (1983) *Alcohol and Industrial Accidents*, Christian Economic and Social Research Foundation, London.

Beaumont, P.B. and Allsop, S.J. (1983) The Beverage Report, *Occupational Safety and Health*, **13**, 25–7.

Beaumont, P.B. and Hyman, J. (1987) The work performance indicators of problem drinking: some British evidence, *Journal of Occupational Behaviour*, **8**, 55–62.

Beiner, L. (1987) Gender differences in the use of substances for coping. In R.C. Barnett, L. Beiner and G.K. Baruch (eds), *Gender and Stress*, New York Free Press.

Blum, T.C., Roman, P.M. and Martin, J.K. (1993) Alcohol consumption and work performance, *Journal of Alcohol Studies*, **54**, 61–70.

Buchanan, D.J. (1988) Studies on blood alcohol in the workers of a Zambian copper mine, *Alcohol and Alcoholism*, **23**, 239–42.

Doogan, K. and Means, R. (1990) *Alcohol and the Workplace*, Occasional Paper 36, University of Bristol, School for Advanced Urban Studies.

Emery, M. (1986) Alcohol and safety in industry, *Journal of the Society of Occupational Medicine*, **36**, 18–23.

Froggatt, P. (1988) *Fourth Report of the Independent Scientific Committee on Smoking and Health*, HMSO.

Glatt, M.M. and Mills, D.R. (1965) Occupational behaviour patterns in samples of English alcoholic employees, *British Journal of Addiction*, **61**, 71–8.

Gomberg, E.S. (1979) Problems with alcohol and other drugs. In E.S. Gomberg and V. Franks (eds), *Gender and Disordered Behaviour*, New York: Brunner/Mazel.

Hertz, R.P. and Emmett, E.A. (1986) Risk factors for occupational hand injury, *Journal of Occupational Medicine*, **28**, 36–41.

Imperial Cancer Research Fund/Cancer Research Campaign (1991) *Passive Smoking – A Health Hazard.*

Institute of Personnel Management (1990) *Smoking Policies at Work – An IPM Guide*, IPM, UK.

Jenkins, M., McEwen, J., Moreton, W.J., East, R., Seymour, L. and Goodin, M. (1989) Smoking policies at work, *Health Education Journal.*

Knupfer, G. (1982) Problems associated with drunkenness in women: some research issues, *Alcohol and Health Monograph*, **4**, Rockville, MD, National Institute of Alcohol Abuse and Alcoholism.

Lee, C. and Gray, J.A. (1994) The role of employee assistance programmes. Cited in C. Cooper and S. Williams (eds), *Creating Healthy Work Organizations*, John Wiley, 215–42.

Lewis, R.J. and Cooper, S.P. (1989) Alcohol, other drugs and fatal work-related injuries, *Journal of Occupational Medicine*, **31**(1), 23–8.

McDonnell, R. and Maynard, A. (1985) Counting the cost of alcohol: gaps in epidemiological knowledge, *Community Medicine*, **7**, 4–17.

Maynard, A. and Godfrey, C. (1994) Alcohol policy – evaluating the options, *British Medical Bulletin*, **50**(1), 221–30.

Maynard, A. and McDonnell, R. (1985) Counting the cost of alcohol: gaps in epidemiological knowledge, *Community Medicine*, **7**, 4–17.

Moll Van Charante, A. and Mulder, P.G.H. (1990) Perceptual activity and the risk of industrial accidents, *American Journal of Epidemiology*, **131**(4), 652–63.

NIAAA (1990) *Alcohol and Health*, Seventh Special Report to US Congress, Rockville, MD, US Department of Health and Human Services.

Powell, M. (1990) *Reducing the Costs of Alcohol in the Workplace*, Discussion Paper 68, Centre for Health Economics, University of York, UK.

Powell, P., Hale, M., Martin and Simon, M. (1971) *2000 Accidents*, London, National Institute of Industrial Psychology.

Robbins, C. (1989) Sex differences in psychosocial consequences of alcohol and drug abuse, *Journal of Health and Social Behaviour*, **30**, 117–30.

Robbins, C.A. and Martin, S.S. (1993) Gender styles of deviance and drinking problems, *Journal of Health and Social Behaviour*, **34**, Dec., 302–21.

Robinson, C.C., Kuller, L.H. and Perper, J. (1988) An epidemiologic study of sudden death at work in an industrial country 1979–1982, *American Journal of Epidemiology*, **128**, 806–20.

Roman, P.M. (1988a) Formal intervention in employee health: comparisons of the nature and structure of employee assistance programs and health promotion programs, *Social Science and Medicine*, **26**(5), 503–14, Pergamon Journals Ltd.

Roman, P.M. (1988b) Growth and transformation in workplace alcoholism programming. In M. Galanter (ed.), *Recent Developments in Alcoholism*, **6**, 131–58, New York: Plenum Press.

Roman, P.M. (1988c) *Women and Alcohol Use: Review of the Research Literature*, Rockville, MD: Alcohol, Drug Abuse and Mental Health Administration.

Roman, P.M. (1990) Alcohol Problem Intervention in the Workplace: Employee Assistance Programs and Strategic Alternatives. New York: Quorum Books.

Roman, P.M. and Blum, T.C. (1989) Alcohol problem intervention in the workplace: data on present status and future implications, *Alcohol Health Research-World*, **13**(4), 375–80.

Sand, E.S. and Madden, J.W. (1976) Certified incapacity and unemployment in alcoholics, *British Journal of Psychiatry*, **128**, 340–5.

Sanders, L.W. (1990) Tobacco control! its problems and profits. In B. Durston and K. Jamrosik (eds), *The Global War, Proceedings of the Seventh World Conference on Tobacco and Health*, Health Department of Western Australia, 753–4.

Scanlon, W.F. (1986) *Alcoholism and Drug Abuse in the Workplace: Employee Assistance Programs*, Praeger, New York.

Smith, A.R. (1986) Alcoholism and gender: patterns of diagnosis and response, *Journal of Drug Issues*, **16**, 407–20.

Sniezek, J.E. and Horiagon, T.M. (1989) Medical examiner reported fatal occupational injuries, North Carolina 1978–1984, *American Journal of Industrial Medicine*, **15**, 669–78.

Sonnenstuhl, W. and Trice, H. (1986) *Strategies for Employee Assistance Programs: The Crucial Balance*, ILR Press, Ithaca, New York.

Trice, H.M. and Roman, P.M. (1972) *Spirits and Demons at Work: Alcohol and Other Drugs on the Job*, Ithaca, New York, Cornell University, New York State School of Industrial and Labour Relations.

Trice, H.M. and Schonrunn, M. (1981) A history of job based alcoholism programs: 1955–1990, *Journal of Drug Issues*, **11**, 171–98.

Webb, G.R., Redman, S., Hennrikus, D., Rostas, J.A.P. and Sanson-Fischer, R.W. (1990) The prevalence and sociodemographic correlates of high risk and problem drinking at an industrial worksite, *British Journal of Addiction*, **85**, 495–507.

Webb, G.R., Redman, S., Hennrikus, D.J., Kelman, G.R., Gibberd, R.W. and Sanson-Fisher, R.W. (1994) The relationship between high-risk and problem drinking and the occurrence of work injuries and related absences, *Journal of Studies on Alcohol*, **55**, 434–46.

White, P. (1987) Tobacco or health, *Smoke Free Europe*, **4**, WHO Regional Office, Denmark.

USEFUL ADDRESSES

Alcohol Concern, 305 Grays Inn Road, London WC1X 8QF.

Alcohol Education and Research Council, Room 6, Abell House, John Islip Street, London SW1P 4LH.

Alcoholics Anonymous, General Service Office, PO Box 1, Stonebow House, Stonebow, York YO1 2NJ.

Action on Smoking and Health, 5–11 Mortimer Street, London W1N 7RH.

Clean Air – Campaign For A Smoke-free Environment, 33 Stillness Road, London SE23 1NG.

Institute of Alcohol Studies, Alliance House, 12 Caxton Street, London SW1H 0QS.

Medical Council on Alcoholism, 1 St Andrew's Place, London NW1 4LB.

Cancer 5

Preamble to workplace exposure to carcinogenic substances • Defining cancer • The role of IARC • Occupational cancer • Asbestos • Asbestos and mesothelioma • Man-made mineral fibres (MMMF) • Rubber • The paper-board industry and renal cancer • Vinyl chloride • Polyvinyl chloride (PVC) • Benzidine and beta-napthylamine • Silica dust • Developments in prevention • Problems associated with education programmes to reduce the risk of cancer • The role of charities in prevention • Examples of good workplace practice • Summary • Suggested further reading • References • Useful addresses • Notes

PREAMBLE TO WORKPLACE EXPOSURE TO CARCINOGENIC SUBSTANCES

After heart disease, cancer is the second largest cause of death per year in the United Kingdom. At least 2–8% of deaths from cancer have occupational associations (Gates, 1995). Globally, the figure is at least 30%.

Occupational exposure has long been recognized as a contributory factor to the development of this condition. Pott, as early as 1775 acknowledged the consequences of exposure to carcinogenic substances in the workplace among chimney sweepers. A substantial number developed cancer of the scrotum as a direct result of their work.

Between 1870 and the early 1900s, soot, coal tar and arsenic compounds were found to be contributory factors in the development of skin and scrotal cancers; this came about as a result of continued exposure.

From the 1920s to the present day, a series of discoveries have been made by researchers relating a variety of compounds and fibres from the work environment to cancer.

Table 5.1 lists some of the common carcinogenic substances found in a variety of workplaces, and is by no means conclusive.

Table 5.1 Common carcinogenic substances found in workplaces

Substance	Cancer: location/type
Arsenic	Lung
Asbestos	Lung, gut, larynx
Auramine	Bladder
4–Aminobiphenyl	Bladder
Benzene	Leukaemia
Benzidine	Bladder
Bis (chloromethyl) ether	Lung
Cadmium	Prostate
Chromates	Lung
Coal tar fumes	Lung
Mineral oils	Skin, scrotum, respiratory tract
Mustard gas	Respiratory tract
2–naphthylamine	Bladder
Nickle compounds	Lung, nose
Radium	Sarcoma
Radon daughters	Lung
Silica dust	Lung, silicosis
Vinyl chloride	Angiosarcoma, lung, brain
Wood dust	Nasal, sinus
X-rays	Skin, leukaemia

DEFINING CANCER

Cancer is a disorder affecting the cells of the body. The disorder begins in a group of cells which, if left untreated, continue to develop and grow into abnormal cells. The rate of growth is dependent on the cancer type. Cancers are able to invade and destroy body tissues as well as enter the blood and lymphatic systems where they are able to spread to other tissues and organs within the body. The resultant formation of 'secondaries' or 'metastases' exacerbate the seriousness of the condition.

PREDISPOSING FACTORS TO THE DEVELOPMENT OF CANCER

An individual's genetic make-up, cultural background, habits and general health status, all determine the likelihood of the development of cancer. There are, however, factors that contribute to the development of cancer. They are:

- **the body's immune system** – a compromised immune defence system increases the likelihood of exposure to possible carcinogenic substances;
- **hormones;**

- **smoking and passive smoking** – as noted in Chapter 4, smoking in whatever capacity can lead to the development of cancer;
- **diet;**
- **environmental pollution;**
- **enzymes;**
- **exposure to industrial chemicals;**
- **medication;**
- **infection** – cancer has an association with the Epstein-Barr virus[1] and the development of the Burkitt tumour,[2] particularly in naso-pharyngeal cancer in the peoples of South China;
- **age;**
- **radiation.**

THE ROLE OF IARC

IARC, the International Agency for Research on Cancer, based in Lyon, is responsible for preparing a series of reports concerning specific substances that are suggested as carcinogenic to humans. Many of these substances are found in manufacturing and production industries.

The agency lists substances such as arsenic, asbestos, benzene, benzidine, sulphate, chromium and vinyl chloride among its substances. A few of these will be discussed later in this chapter.

OCCUPATIONAL CANCER

There is increasing evidence of an association between types of occupation, working with specific substances and the development of cancer. Substances to which the employee is exposed in the course of normal duties, can be absorbed or inhaled and formed during metabolism within the body. Alternatively, they may precipitate additional substances or agents. What is clear is that occupational cancer does not simply arise as a result of exposure to one single carcinogenic substance.

During the earlier part of this century, researchers recorded a person's genetic characteristics such as skin pigmentation, to assist in determining whether or not they would develop skin cancer. This was noted following exposure to ultraviolet light. Concrete evidence relating to multi-factoral causation occurred in 1927 when researchers demonstrated that skin cancer was more likely to develop if the skin was scratched before exposing it to tar. This has also been demonstrated with asbestos and uranium. Cohorts of asbestos and uranium miners showed much higher rates of cancer development when they also smoked than those who did not.

There are, however, some problems in identifying the genuine causative agents. The reasons are as follows.

1. There is frequently a mismatch in the time between exposure and effect upon individuals. Any manifestation of cancer occurs a considerable time after exposure. Indeed, the effects of exposure to carcinogenic substances are irreversible. Benzidine, for example, is known to lead to the development of cancer some 30 years after initial exposure. Substances such as PCB, ethylene dichloride and vinyl chloride also stay in the body for many years following initial exposure.
2. Some diseases such as lung cancer are relatively common among the population, so it is often problematic to detect incidences of exposure among specific occupational groups. Factors such as diet, smoking, lifestyle or separate exposure to other substances may act harmoniously with other workplace factors to affect the incidence of cancer.
3. The actual cause of death may not be attributed to exposure to particular carcinogenic substances.
4. Recognition that health problems in the workplace are apparent. Often it requires a significant body count and sophisticated studies to identify an excess exposure to substances deemed to be carcinogenic.

As indicated in Table 5.2, there are a variety of industries and substances within the workplace to which individuals may be exposed. Developments in research alert us to new compounds on a regular basis as causative agents of cancer.

In view of the extent of the subject, it is intended to base discussions on current development in the field of carcinogenic exposure to selective substances in the workplace and the ways in which issues that arise for employees have been tackled. The emphasis, as throughout this text, is on practicalities, prevention and guidance for improved practice.

ASBESTOS

In the last few months (1996), a British court overturned a claim by the company T & N for its exposure of children during the 1930s to asbestos fibres outside its plant works. Although one of those suffering as a result of exposure has since died as a result of cancer, the court awarded a total of £150 000 to the victims as a direct result of the company's negligence. The court claimed that evidence has been available since 1925 of the dangers of asbestos, so the company should have taken steps to protect its workforce and the public.

Asbestos is a fibrous mineral having tremendous heat-binding and chemical resistance properties. It is also a chemical compound known as silicate. Asbestos[3] is used in a variety of ways, but the most popular uses are for curtains, shields, clutch facings, pipe coverings and reinforcement for plastics and cements.

Table 5.2 Industries and occupations associated with increased risk of the development of cancer

Industry	Occupation
Agriculture, forestry, fishing	Application of arsenic in vineyards
Mining	Arsenic, asbestos, uranium and haematite mining
Asbestos production	Insulating material production, e.g. pipes, metal sheeting, clothes, masks, asbestos, cement production
Shipbuilding and transport	Dockyard work
Chemical	Bis (chloromethyl) ether and chloromethyl methyl ether (BOME and CMME) production and use. Vinyl chloride producers. Isopropyl chromate production, dye manufacture and use, auramine manufacture
Pesticides and herbicides	Arsenical insecticides production and packaging
Gas industry	Coke plant and gas work
Rubber industry	Rubber manufacture, milling, mixing, synthetic latex production, tyre curing, cable making
Construction	Insulation pipe covering
Leather	Boot and shoe manufacture/repair
Petroleum	Shale oil production
Metal	Copper smelting, chromate production, nickel refining, ferrochromium production, metal machining
Wood pulp and paper production	Furniture and cabinet making
Textiles	Mule spinning

In a case study reported in 1907, Montague Murray was the first to note the adverse effects related to working with asbestos. It was originally labelled 'asbestos poisoning' and later asbestosis. The symptoms associated with the disease included coughing, emaciation, dyspnoea, weakness and chest pain.

Cook (1927) and McDonald (1927) were the first to describe 'curious bodies' in the lungs of men and women who had died of asbestosis.

However, it was not until 1935 that Lynch and Smith reported the first case of carcinoma of the lung in a patient with asbestosis. Throughout the late 1930s and early 1940s, other articles were published discussing the occurrence of lung cancer in patients with asbestosis. It was 1941 when Nordmann and Sorge published their findings of animal experiments that resulted in the production of pulmonary cancer in 20% of the surviving animals who had been exposed to asbestos.

By 1949, the association between asbestosis and lung cancer had been firmly established in America. Further evidence came from Doll (1955) who carried out an extensive epidemiological study, wherein he estimated the risk of lung cancer in asbestos workers to be ten times greater than the average employee. In addition to these findings, asbestos inhalation has since become associated with cancers of the stomach, oesophagus, colon, rectum, larynx, pharynx and peritoneum.

ASBESTOS AND MESOTHELIOMA

A report in *The Lancet* (Peto *et al.*, 1995) predicted that male deaths from mesothelioma[4] could peak around the year 2020, reaching between 2700 and 3300 deaths within that year. At present the level is around 1000.

Current concerns exist surrounding the death rate of men under 50 years old, the majority of whom started work in the mid-1960s, which are likely to increase in years to come. The report identified that the groups of workers most at risk are plumbers, carpenters, gas fitters, electricians and construction workers.

Current legislation exists to protect employees as far as is reasonably practicable from the adverse effects of exposure to potential or known carcinogenic substances. They include:

- The Asbestos (licensing) Regulations 1983
- The Control of Asbestos at Work Regulations 1987 (amended 1992)
- The Asbestos (Prohibitions)) Regulations 1992.

MAN-MADE MINERAL FIBRES (MMMF)

It was 1972 when the first evidence arrived concerning the hazards of fibrous glass and other man-made mineral fibres. In 1979, a group of trades union officials made a decision to treat MMMF with the same caution as asbestos. This was as a result of experimental evidence that glass fibres lead to the development of mesothelioma.

By 1980, a conference on fibrous dust highlighted the need for a prevention programme and the control of dust emission into the environment. It later transpired that investigations in North America and Europe detected that almost 8000 deaths in a labour force of 41 000 production and maintenance workers was a result of MMMF (Doll, 1987).

In 1982, preliminary evidence from the European sponsored investigation was presented to the WHO Conference. Further conclusions drawn from observations in Europe and North America were as follows:

1. A risk of lung cancer in employees in the rock and glass wool sectors of the MMMF industry amounted to 25% above the norm, 30 years after initial employment.
2. The risks were greater in the rock sector.

To date, the evidence is inconclusive for glass wool and other MMMF, but it will be interesting to note any future developments.

RUBBER

It was during the early 1930s that rubber was reported as a likely carcinogenic substance in Britain. This was highlighted in employees in the rubber industry having a higher than average death rate from cancer.

Further developments in the 1950s saw that rubber workers had a rate of death from cancer at, on average, 8% higher than workers in other manufacturing industries. The most common cancers were those of the large bowel, respiratory system, lymph glands and the bloodstream.

This prompted a series of studies into the carcinogenic capacity of rubber. One of the largest was undertaken in the 1970s by the United Rubber Workers and four rubber companies – Firestone, Goodyear, General Tire & Rubber and UniRoyal – with scientists from the University of North Carolina School of Public Health. This ten-year study found that a higher than expected number of deaths from cancer in general and cancers of the stomach, colon, prostate, lymph and bloodstreams developed in the majority of employees of all four companies.

Similar studies were undertaken in Britain and the British Rubber Manufacturers Association noted that:

- there was an absence of bladder cancer in workers who started work after 1951;
- stomach cancers (29%) had developed in workers employed in the making of tyre and rubber goods (in addition, lung cancer (23%) and other cancers of the oesophagus and intestines had also developed).

To date, the International Agency for Research on Cancer (IARC) has collected evidence from a variety of sources of employees in Asia,

America and Scandinavian rubber industries. So far it has arrived at the following conclusions.

- Increased risk has been found from leukaemia and cancers of the bladder, lung and stomach.
- There is isolated evidence of cancers of the brain, pancreas, thyroid and oesophagus as a result of exposure to rubber.
- In one particular study, the prevalence of increased risk of exposure was greater in women than in men.

In Britain, claims for compensation are currently being made against rubber manufacturers. These cases are now coming forward as a result of employees' initial exposure to carcinogenic substances as far back as 1949. It is expected that under the COSHH Approved Code of Practice 1994, claims will be reduced.

THE PAPERBOARD INDUSTRY AND RENAL CANCER

In June 1986, the National Institute for Occupational Safety and Health (NIOSH) received a request for technical support from the Occupational Safety and Health Administration to investigate a possible cancer cluster at a paperboard packaging manufacturing plant.

The plant produced printed paperboard boxes for use as food containers. The problem was highlighted when a local physician had reported that one worker had been diagnosed with renal cell cancer and another with bladder cancer whilst working in the finishing department. Three potential carcinogenic substances were identified following a review of the company's safety data sheets. The substances were methylene chloride, formaldehyde and trichloroethylene. The main causative substance predisposing individuals in the paper drying areas of the pulp and paper mill industry was formaldehyde (Sinks *et al.*, 1992).

The study discussed above represented the first reporting of excess renal cell cancer among paperboard print manufacture employees. The excess risk was found to be associated with length of employment in the organization rather than limited to any single department or work process. There is further need for research into this phenomenon as it is not yet clear why excess risk of renal cell cancer occurs among paperboard printing workers.

VINYL CHLORIDE

Vinyl chloride has been identified as a causative agent of cancer (IARC, 1979). Although vinyl chloride was initially reported in the UK to produce tumours in rats, measures to reduce levels of exposure of workers were not taken until 1974, following a report of three cases of the rare angiosarcoma[5] of the liver in extremely exposed workers. Vinyl chloride is also known to cause cancers of the brain, lung, lymphatic tissue and blood.

The harmful properties of vinyl chloride have been in the public domain since the 1940s, when vinyl chloride gas was utilized for medical anaesthesia. It was swiftly abandoned when it was discovered to cause cardiac arrhythmias.

It was during the 1960s that a discovery was made relating to further problems with vinyl chloride association. The condition acroosteolysis developed as a result of employee exposure during the process of polymerization to polyvinyl chloride.

The clinical manifestations of the disease are tenderness of the finger tips later accompanied by the gradual destruction of the bones of the fingers. The blood vessels were also affected, leaving affected workers with a condition similar to Renaud's disease.

It was during 1971 that Viola *et al.* recognized the cancerous effect of vinyl chloride. This was later confirmed in 1976 by the US Goodrich Company, a manufacturer of PVC, which announced to its employees, NIOSH, and the Kentucky Department of Labour, that three of its workforce had died of angiosarcoma of the liver. The tumours occurred between 14 to 27 years after exposure. This was followed by other producers of vinyl chloride and polyvinyl chloride making similar claims. Experiments such as Maltoni's (1976) followed confirming the production of vinyl chloride as cancerous.

A variety of control measures have been developed to make this product safer. However, the debate on vinyl chloride and its byproduct polyvinyl chloride continues.

POLYVINYL CHLORIDE (PVC)

The main use of vinyl chloride is for the production of PVC which is in turn used to manufacture plastic materials. It is known to lead to the greatest risk of exposure when vinyl chloride is polymerized to form PVC resins.

This substance was patented in 1913 and is regarded as a major consumer of chlorine. It was established in Germany in the 1930s as a prominent industrial commodity. It achieved considerable success in the market during the 1960s and is still produced in great quantities within Europe today.

This versatile plastic product used in a variety of applications has developed cause for concern among environmental pressure groups.

Greenpeace International makes the claim that PVC poses an unacceptable environmental risk throughout its production, use and disposal. The risk arises from the production process.

The elements involved in PVC production are ethylene and chlorine. When combined they form ethylene dichloride. The product is then converted to vinyl chloride monomer (VCM). Approximately one-third of the bi-product generated through VCM production is converted into a range of dry-cleaning fluids and toilet disinfectants. The remaining waste product is incinerated, injected into deep wells or landfilled.

Both ethylene dichloride and VCM are flammable, explosive and carcinogenic. Long-term exposure to high concentrations at work can lead to cancer of the liver. It has also been associated with 'sick building syndrome'[6] in Sweden.

As a result, a number of countries are imposing voluntary environmental controls on the substance. Indeed, the European Council for Vinyl Manufacturers (ECVM) in 1995 launched its charter at a conference in Brussels, compelling PVC manufacturers to conform to agreed environmental standards.

It is too soon to evaluate the outcome of this request.

BENZIDINE AND BETA-NAPTHYLAMINE

Rehn in 1895 discovered that workers developed bladder tumours as a result of exposure to aniline, a substance derived from benzidine. Since then, a number of studies have linked beta-napthylamine and benzidine to genotoxicity, immune suppression and bladder cancer.

Dye production workers in the US, China, Japan and Britain have shown increased bladder cancer risk as a result of occupational exposure to benzidine and beta-napthylamine. Benzidine is used in security printing, the chemical detection of blood and in aniline dyes. The main route of entry to the body following exposure occurs by absorption through the skin.

A study in Moscow recently showed that exposure to beta-napthylamine and benzidine conclusively led to the development of cancer amongst its workforce. This was the first study of its kind to present gender specific relative rate estimates for women who appeared to have greater or similar bladder cancer rate estimates than men. In addition, the rate was elevated among workers with less than 3 years of benzidine exposure. It suggests that despite controls and prevention programmes, the precautions taken by workers were deemed to be ineffective.

(Bulbulyan *et al.*, 1995)

SILICA DUST

To date silica dust exposure amongst workers has been identified in occupations relating to slate quarrying, sandblasting, tunnelling and refactory brick production.

One of the most significant studies carried out by Merlo *et al.* (1990) demonstrated an increased likelihood of contracting cancer and diseases of the respiratory system following excessive exposure to silica dust in the workplace. The historical mortality study was conducted in Italy on patients treated at San Martino Hospital, Genoa, between 1961 and 1980. The team were alerted to the problem by the hospital's occupational health department following the detection of similarity in characteristics of those patients entering hospital for treatment of respiratory problems.

Silica dust exposure has also taken its toll on Japanese workers. A significantly high ratio for lung cancer was observed in a cohort of 3335 hospitalized pneumoconiosis[7] patients from 11 hospitals located throughout Japan. Of the types of pneumoconiosis identified, silicosis[8] was found to be associated with the highest rate of lung cancer. The individuals identified in the study had never smoked.

IARC are continuing to conduct studies to discover the extent of carcinogeniety of silica dust and its association with the development of lung cancer. To date, the potential carcinogenic effects of silica dust are characterized by three main findings.

1. Silica, as identified by Chiyontani *et al.* (1990), has the potentially harmful effect of being carcinogenic.
2. Lung cancer risk is increased among workers exposed to silica and not exclusively among those exposed to known carcinogens.
3. Those workers who develop silicosis are at an increased risk of developing lung cancer.

DEVELOPMENTS IN PREVENTION

EUROPEAN CANCER CODE

In 1986, the heads of state of the European Community launched the European Programme Against Cancer. The programme arose as a direct response to the need to reduce the risks of developing cancers. Its ultimate goal is to reduce the number of deaths caused by cancer to 15% by the year 2000.

The programme has two main themes:

1. providing information to the public;
2. training members of the various health professions to educate the public and improve their practice in promoting health.

Public information has arisen in the form of leaflets and posters highlighting the 10 Point European Code Against Cancer.

The Code advises:

1. not to smoke – smokers are to give up and are not to smoke in the presence of others;
2. to take alcoholic drinks in moderation;
3. to avoid exposure to the sun;
4. to adhere to health and safety instructions;
5. to eat plenty of fresh fruit, vegetables and foods rich in fibre;
6. to avoid becoming overweight;
7. to visit a doctor if a lump or mole on the body changes or begins to bleed;
8. to visit a doctor if problems arise relating to inexplicable weight loss, hoarseness of voice and bowel habit changes;
9. women to have regular cervical smears;
10. women should examine their breasts on a regular basis.

(Adaptation of the 10 Point Code)

The development of the programme in the UK began in 1989. Contribution of a £1 million grant was provided to fund the project from the Department of Health and the European Commission. Activities were co-ordinated by the Cancer Education Co-ordinating Group (CECG). Projects and activities since its inception have included 'Europe Against Cancer Week'. The programme links with the government's Health of the Nation Strategy, outlining specific targets for cancer reduction. They are:

- to reduce deaths from breast cancer by 25% by the year 2000;
- to reduce deaths from cervical cancer by 25% by the year 2000;
- to halt the year-on-year rise in melanoma by the year 2005;
- to reduce deaths from smoking by adults by 20% by the year 2000.

PROBLEMS ASSOCIATED WITH EDUCATION PROGRAMMES TO REDUCE THE RISK OF CANCER

Box and Anderson's (1993) research into cancer education programmes found that professionals and lay people alike have some difficulty in accepting the prevention message. This arises because they have attitudes and beliefs which have been shaped by their past experiences of cancer.

CANCER WEEK 1992

> This was a predetermined strategy by the European Commission for all Nation States. The theme was 'Work at Your Health', a campaign to promote awareness of the EAC's 10 Point Code in the workplace. The programme's link with the European Year of Health and Safety was geared towards encouraging workplaces to develop positive health promotion programmes including cancer prevention and education. Information packs were produced and disseminated with support and guidance from health promotion departments. Two years later, the DoH, EAC, CECG and Cancerlink, with Ciba Pharmaceuticals launched an exhibition 'All About Cancer'. This was developed following a needs assessment exercise in various workplaces by health promotion officers and personnel managers. The projects have been reasonably successful.

Perceptions relating to anxiety, pain, fear and death are often, according to the researchers, barriers to accepting the prevention message. Lifestyle and behaviour change are long-term strategies and thus may limit the amount a professional may be able to deal with in their capacity as specialists in prevention.

Box and Anderson recommend that training for cancer prevention should involve addressing individual's preconceptions and personal agendas before a way forward can be developed. Allied to this fact, they advise that any education and training programme for cancer prevention should contain:

- exploration of values and beliefs;
- understanding of research underpinning issues relating to cancer;
- techniques to deal with ethics of cancer prevention, education, planning, implementation and evaluation.

THE ROLE OF CHARITIES IN PREVENTION

The main involvement of charities in the prevention of cancer includes:

1. research into treatment and palliative care;
2. public information and education;
3. prevention, early diagnosis and screening;
4. training for all professional health groups;
5. counselling and support for patients and families.

Charities have also made their contribution in an advisory capacity to the Europe Against Cancer initiative.

Some organizations have steered their involvement into more specific directions:

- **Ulster Cancer Foundation** – this foundation has been instrumental in promoting issues relating to smoking prevention among teenagers and cancer prevention programmes in the workplace.
- **The Imperial Cancer Research Fund Unit** – the majority of their work and research underpins the national UK strategy for cancer prevention to the work of Doll and Peto.
- **Marie Curie Care** – this charity organizes and implements cervical and breast screening courses for general practice nurses and has also included skin cancers and malignant melanoma in its programme. This organization also vigorously promoted the EAC 10 Point Code Against Cancer.

EXAMPLES OF GOOD WORKPLACE PRACTICE

THE NATIONAL HEALTH SERVICE, UK

The National Health Service (NHS) Breast Screening Programme's main aim is to raise the profile of breast screening against cancer and early detection. The programme has concentrated its efforts also into the psychological and sociological factors affecting attendance, diagnosis and follow-up treatment. The programme co-ordinators and practitioners liaise closely with GPs and practice nurses, and provide them with education surrounding the service and prevention strategies.

ICL, UK

ICL is a large manufacturing company containing approximately 2500 employees. Its occupational health service took part in a Health Education Authority pilot scheme to raise awareness of cancer issues during 1992. Despite initial resistance, issues relating to care of the skin, diet and smoking cessation were introduced to the workforce as a preamble to the topic of cancer.

Cancer awareness information packs were distributed to all employees at the site. The packs contained guidance for self-examination of the skin, breasts and testicles. A breast-screening service was also offered to all female staff and the wives of all its ICL employees by a consultant radiologist. Group advisory sessions were held by the occupational health nurses to discuss issues relating to cancer by the HEA adviser.

The nursing team also received training on cancer counselling skills.

The company employers also later made requests to back-up services including GP, local hospital and voluntary groups such as Macmillan Nurses. The initiative was a success because of teamwork, managerial support and efficient dissemination of information to suite the needs of the company's employees.

TELECOM, AUSTRALIA

The **Cover Yourself Against Skin Cancer** campaign was undertaken by Telecom workers in Australia. The campaign proved to be an effective tool for getting outdoor workers to increase their sun protection.

Excessive exposure to ultra violet rays from the sun is a major risk factor in the development of skin cancer. The greatest risk occurs during the summer months, as radiation is at its peak between 11.00 am and 3.00 pm in Australia.

Borland *et al.* (1991) evaluated an education package developed by a marketing firm for employees of the company. The package consisted of a video depicting the consequences of malignant melanoma and information to encourage skin protection. The information included keeping a shirt on, wearing a hat, wearing sunscreen and avoiding exposure to the sun at the major risk times.

The result was improved protection of employees by up to 10%, a success for a relatively small-scale campaign.

WELL WORKS PROJECT, USA

The Well Works Project was conducted across 24 organizations during the early 1990s employing over 16 000 employees in Eastern and Central Massachusetts over a two to two and a half year period.

The project was aimed at both individual and organizational alterations relating to smoking reduction, decreased dietary fat consumption, increased dietary fibre intake and reduction of the potential for exposure to occupational carcinogenic substances.

Well Works is part of a long-term research project funded by the National Cancer Institute.

The companies involved were selected from a Dunn and Bradstreet listing and selection was based on:

- number of workers up to 2500;
- staff turnover of less than 20% per year;
- non-English-speaking employees, less than 20%;
- known or suspected use of occupational carcinogenic substances.

The overall purpose of the project is to integrate health protection and health promotion which includes:

- joint employee and manager participation in planning and implementation of the programme. This is vital for gaining authority, promoting environmental change and for employees to take some responsibility over their lifestyle;
- consultation relating to workplace alterations;
- co-ordinated educational programmes targeting health behaviour change.

Within the project, it was necessary to set up Employee Advisory Boards (EAB) to advise and participate in planning, promotion and implementation of programmes. In addition, this board serves as a social marketing function by informing project staff of employees needs and expectations, as well as conveying the programme advisors' expectations and messages. The EAB liaise closely with the Well Works team to ensure smooth running of initiatives.

Well Works is reliant on the hierarchy of controls model for reduction of exposure to workplace hazards. The crux of this model is control of the hazardous substance as close to the source as possible. A series of control measures are introduced depending on the extent to which it is or has been problematic to the organization. The initial measure includes substitution of actual/potential carcinogenic substances for safer ones. This is followed by the use of engineering controls such as ventilation systems which provide protection without being reliant on individual action. Job redesign, job rotation and other administrative changes may also be implemented as further means of reducing potential exposures. Protective personal equipment is regarded as the last resort as reliability, predictability and effectiveness can alter.

Within the project, organizational and employee interventions were made.

(a) Organizational

A review of the companies' Material Safety Data Sheets (MSDS) was carried out. The sheets carried descriptions of substances used in the workplace. In 22 of the 24 organizations studied, the companies' industrial hygienist reviewed 14 630 MSDS, 16% of which listed known or suspected carcinogenic substances. The industrial hygienist also performed a walk-through assessment on each site.

Recommendations were then made outlining specific requirements for each company. These were highlighted as topics for discussion at each EAB meeting.

The majority of companies exhibited similar difficulties, and found common solutions:

1. the need to update their MSDS;
2. to develop methods to review toxicity data on chemicals prior to their introduction to the workplace;
3. to review the roles and functions of their health and safety committees;
4. to provide information on cancer prevention for their employees;
5. to explore legislative requirements.

(b) Employees

The Well Works Project includes advice and solutions relating to lifestyle and employee protection, that is relating to smoking reduction, dietary help and prevention against excessive exposure to hazardous substances. The companies involved in the project permitted the attendance of their employees to educational activities based on a 'learning by doing' model, on a regular basis for 15–20 minutes on each occasion.

On the whole, the project has assisted in the reduction of actual/ potential carcinogenic substances entering the organizations involved. Employees and managers have gained a more coherent approach as supportive and connected teams in the drive towards cancer prevention.

SUMMARY

The risk of cancer can arise from a multiplicity of substances, environmental conditions and factors beyond human control.

The potential for exposure to carcinogenic substances in the workplace is extremely high, particularly if employers are not kept abreast of current legislation and managers do not make their teams aware of the risks and how to prevent them. Examples can be taken from the various workplaces mentioned in this chapter to enable proactive managers to improve and protect worker health.

Most of all:

- identify the potential carcinogenic substances;
- ensure everyone in the organization is aware of any possible dangers to hazardous substances and provide means of protection;
- check substance emissions, safety in packaging and storage procedures regularly;
- record any accidents in company documents;
- keep abreast of changes in legislation;
- consult experts if uncertain of likely problems the substance(s) may cause;
- support and involve workforce in awareness-raising initiatives.

SUGGESTED FURTHER READING

Anderson, Y. and Weston, D. (1993) *Work at Your Health, CECG and EC*, Ridgeway Training, Swindon, Wiltshire.

Egilman, D.S. and Reinert, A.A. (1995) The origin and development of the asbestos threshold limit value: scientific indifference and corporate influence, *International Health Service Journal*, **25**(4), Baywood Publishing, 667–96.

Health Education Authority (1992) *Cancer at the Workplace*, HEA, London.

Heller, T., Bailey, I. and Pattinson, S. (1992) *Preventing Cancers*, Open University Press.

Heller, T., Davey, B. and Bailey, L. (1992) *Reducing the Risk of Cancers*, Hodder & Stoughton in Association with the Open University and the European Commission.

HSE and EAC (1993) *Health and Safety Executive Conference Proceedings 1993, on Carcinogens at the Workplace*, HSE, UK.

REFERENCES

Borland, R.M., Hocking, B., Godkin, G.A., Gibbs, A.F. and Hill, D.J. (1991) The impact of a skin cancer control education package for outdoor workers, *The Medical Journal of Australia*, **154**, 20 May.

Box, V. and Anderson, Y. (1993) *Professional Attitudes to Cancer Survey*, University of Southampton.

Bulbulyan, M.A., Figgs, L.W., Zahm, S.H., Savitskaya, T., Goldfarb, A., Astashevsky, S. and Zaridze, D. (1995) Cancer incidence and mortality among beta-napthylamine and benzidine dye workers in Moscow, *International Journal of Epidemiology*, **24**(2), 66–275.

Chiyontani, K., Saito, K., Okubo, T. and Takahashi, K. (1990) Lung cancer risk among pneumoconiosis patients in Japan with special reference to silicosis. In L. Simonato *et al.* (eds), *Occupational Exposure To Silica and Cancer Risk*, IARC.

Cook, W.E. (1927) Pulmonary asbestosis, *British Medical Journal*, **2**, 1024–5.

Creech, J.L. Jnr, Johnson, M.N. (1974) Angiosarcoma of the liver in the manufacture of polyvinyl chloride, *Journal of Occupational Medicine*, **16**, 150–1.

Doll, R. (1955) Motality from lung cancer in asbestos workers, *British Journal of Industrial Medicine*, **12**, 81–6.

Doll, R. (1987) Overview and conclusions: symposium on man made mineral fibres, Copenhagen, October 1986, *Annals of Occupational Hygiene*, **31**, 805–19.

Gates, E. (1995) Occupational cancer, *Health and Safety at Work*, November.

IARC (1979) *IARC Monographs on the Evaluation of the Carcinogenic Risk of Chemicals to Humans, Vol.19, Some Monomers, Plastics and Synthetic Elastomers and Acrolein*, International Agency for Research on Cancer, Lyon.

Lynch, K.M. and Smith, V.A. (1935) Pulmonary asbestos III: carcinoma of lung in asbestos-silicosis, *American Journal of Cancer*, **25**, 56–64.

McDonald, S. (1927) Histology of pulmonary asbestosis, *British Medical Journal*, **2**, 1025–6.

Merlo, F., Doria, M., Fontana, L., Ceppi, M., Chesi, E. and Santi, L. (1990) Mortality from specific causes among silicotic subjects: a historical prospective study. In L. Simonato, A.C. Fletcher, R. Saracci and T.L. Thomas (eds),

Occupational Exposure to Silica and Cancer Risk, Lyon, International Agency for Research on Cancer, IARC, 1990.

Peto, J., Hodgson, L., Matthews, F. and Ross, J. (1995) Continuing increase in mesothelioma mortality in Britain, *The Lancet*, **345**(4), March, 535–9.

Pott, P. (1775) *Chirurgical Observations Relative to the Cataract, the Polypus of the Nose, the Cancer of the Scrotum, the Different Kinds of Ruptures and the Mortification of the Toes and Feet*, Clarke and Collins, London, 63–5.

Sinks, T., Lushniak, B., Haussler, B.J., Sniezek, J., Deng, J.F., Roper, P., Dill, P. and Coates, R. (1992) Renal cell cancer among paper board printing workers, *Epidemiology*, **3**, 483–9.

Viola, P.L., Bigotti, A. and Caputo, A. (1971) Oncogenic Response of Rat Skin, Lungs and Bones to Vinyl Chloride, *Cancer Research*, **31**, 516–22.

Xue-Yun, Y., Ji-Gang, C. and Yong-Ning, Hu. (1990) Studies on the relation between bladder cancer and benzidine, or its derived dyes in Shanghai, *British Journal of Industrial Medicine*, **47**, 544–52.

USEFUL ADDRESSES

British Association of Cancer United Patients (BACUP), 3 Bath Place, Rivington Street, London EC2A 3JR.

Cancerlink, 17 Britannia Street, London WC1Z 9JN.

Cancer Relief MacMillan Fund (CRMRF), 15–19 Britten Street, London SW3 3TZ.

Marie Curie Cancer Care, 28 Belgrave Square, London SW1 8QG.

Imperial Cancer Research Fund, PO Box 123, Lincoln's Inn Fields, London WC2A 3PX.

NOTES

1. The Epstein-Barr Virus, named after the British pathologist, M.A. Epstein-Barr, is a virus which is thought to be the causative organizm of infectious glandular fever.
2. Burkitt's Tumour, named by the Irish surgeon, D.P. Burkitt, is also known as African Lymphoma. It is a lympho-sarcoma (cancer arising from the lymph gland) occurring almost exclusively in children living in areas of Central Africa. It is known to attack the jaw, lymph nodes, kidneys and thyroid gland.
3. Asbestos is known in industry by other names including: actinolite, amianthus, amosite, amphibole, anthophyllite, ascarite, chrysotile, crocidolite and tremolite.
4. Mesothelioma is a cancer affecting mainly the pleura and peritoneum of the lungs. It is a cancer known to increase in its development some 15 years following exposure. In the UK, the Health and Safety Executive has recorded mesothelioma as a cause of death. It has risen from 154 cases in 1968 to 1009 in 1991.
5. Angiosarcoma is a malignant vascular growth.
6. Sick building syndrome is the name given to the conditions that arise as a result of a poorly ventilated and poorly lit building. Particularly common in

large office blocks, workers exhibit lethargy, eye strain, sore eyes, headaches and generalized irritation.

7. Pneumoconiosis refers to fibrosis of the lungs due to inhalation of dust particles.
8. Silicosis is fibrosis of the lungs following inhalation of silica dust particles; it is usually accompanied by tuberculosis.

HIV and AIDS 6

Introduction • Factors concerning transmission • Cases of occupational transmission • Concerns for the healthcare worker • The food industry • Employment, law and HIV • The testing debate • Policy and the workplace • Education and training • Summary • Suggested further reading • References • Useful addresses

INTRODUCTION

This chapter is written on the premise of the need to emphasize the problems and difficulties that managers and employees need to tackle in the control of transmission of HIV infection and to reduce the stigma of AIDS, should an employee wish to disclose his or her condition. Although there is a plethora of information surrounding the conditions from a variety of perspectives, the intention for the most part is to address issues specific to the workplace.

The WHO estimates that by the year 2000, global statistics for HIV will reach between 30 and 40 million. Over two-thirds of companies with between 2500 and 5000 employees and almost one in twelve companies with fewer than 500 employees in the US have had an employee with HIV/AIDS.

Recorded workplace cases to date in the UK have been patchy and relate mainly to occupational transmission as a direct result of working in healthcare or laboratory employment.

> 1634 cases of AIDS were reported in England in 1994; these brought the cumulative total of reported AIDS cases since 1982, to 9510, of whom 6443 are known to have died. It is estimated that on average, people newly diagnosed with AIDS were infected with HIV some ten years previously.

THE VIRUS

The human immuno deficiency virus (HIV) is transmitted from an infected person to another either by sexual intercourse or by exchange of blood. Once a person has HIV it can take up to six months before the antibody is produced. The presence of the antibody can be tested for and it is the detection of the antibody which reveals that the person has HIV. Infection by the virus makes a person infectious to others though s/he may look and feel healthy.

Once a person is infected, a spectrum of possibilities emerges:

1. The person may remain 'antibody positive' and show no other signs. This may continue for years.
2. At any time from being antibody positive, the person may develop a serious indisposition called AIDS-related complex (ARC). This is caused by HIV but is not AIDS in its full sense.

   ARC manifests itself through a collection of symptoms: great fatigue, constant fevers, inexplicable weight loss, persistent diarrhoea, night sweats, shingles, herpes and thrush in the mouth. This condition can be episodic and the individual may have periods of reasonably good health, though always infected with HIV.
3. A less serious condition may also develop known as generalized lymphadenopathy. Here the glands in the neck and armpits become persistently swollen. This too is likely to recede and return at intervals.
4. In some cases the virus attacks the brain and causes dementia. The individual may become confused, moody, unable to make decisions and forgetful.
5. Finally, the virus may lead to AIDS itself. It can do this either directly or indirectly via episodes of ARC or lymphadenopathy. The condition leads to severe defects in many cells which are a crucial aspect of the body's immune system. The defect leaves the person susceptible to infection by certain organisms which cause illnesses that rarely infect individuals. The first is a skin cancer known as Kaposi's Sarcoma; the second is a type of pneumonia known as pneumocystis carnii.

FACTORS CONCERNING TRANSMISSION

HIV has been located in semen, tears, blood, breast milk, cervical secretions, cerebrospinal fluid and saliva. However, only three of these seem able to pass the virus to an uninfected person: semen, cervical secretions and blood.

Sexual transmission

- *Between heterosexuals*: occurs through vaginal or anal intercourse. The virus therefore can be passed from an infected woman to an uninfected man or from an infected man to an uninfected woman. The body fluids involved are semen and cervical secretions.
- *Between homosexual men*: the infection can pass from the inserter to the receptive partner or in the opposite direction. The body fluids playing a part are semen and possibly blood.

Blood transmission

This can occur through the following.

- Transfusion – if a person receives a transfusion of infected blood or blood products they are at risk of contracting the virus. Individuals requiring regular blood transfusions, such as haemophiliacs or those with sickle cell anaemia, are at greatest risk if the blood/product is untreated. Transfusion transmission is considered to be extremely rare as there are stringent regulations and mechanisms employed by blood transfusion services to prevent this occurring.
- Injection – if the contaminated blood is in a needle or syringe. This can occur if people inject drugs for recreational use, or if addicts share a needle, syringe or mixing bowl. Care workers are now considered to be at risk. (See section on care workers later in this chapter).

CASES OF OCCUPATIONAL TRANSMISSION

Transmission of HIV infection has occurred mainly among healthcare and laboratory employees.

There are, to date, approximately 182 cases worldwide of transmission of HIV in the workplace. The majority of documented cases appear to have resulted from contact with patients' infected blood or body fluids containing blood.

Although the number of incidences of occupational transmission of HIV is currently unknown, reporting has improved since the mid-1980s. In the UK the Public Health Laboratory Service (PHLS) is responsible for maintaining a register with details of any outcomes following accidental exposure. The European Commission has also set up a multicentre study of occupational exposure to HIV with surveillance measures taken in 13 European countries. To date, the most commonly reported incidences are as follows.

- Disposal of needles, particularly when re-sheathed after use. Studies have shown that up to 60% of needlestick injuries have occurred as a result of re-sheathing needles. In fact, the estimated rate of transmission of HIV needlestick injury infection is 1: 275; when compared to hepatitis B this is rather low. Hepatitis B's estimated rate of transmission is 1: 20.

- The skin being penetrated or torn by sharp instruments such as blades and needles that have been used on known/unknown HIV positive patients.
- Mucocutaneous exposure such as contact with torn skin and mucous membranes of the eyes, nose or mouth, also prolonged exposure to large quantities of infected body fluids.

CONCERNS FOR THE HEALTHCARE WORKER

There are increasing concerns for healthcare workers (HCW) regarding the risk of HIV infection in the workplace (O'Donnell *et al.*, 1987; van Servellen *et al.*, 1988; Wallack, 1989; Pomerance and Sheilds, 1989; Symser *et al.*, 1990; Currey *et al.*, 1990; Feldman *et al.*, 1990; Scherer *et al.*, 1992; Breault and Polifroni, 1992; van Wissen and Siebers, 1993).

Even though, as mentioned earlier, the risk is considerably lower than contracting hepatitis B, the fear of infection, nonetheless, persists. During 1993, there were at least 60 documented cases of HCWs' becoming antibody positive following exposure to HIV infected blood. At least between 200 and 300 HCWs die each year as a direct or indirect result of occupationally acquired hepatitis B. As many as 12 000 become infected with the virus. Little concern has been voiced concerning this fact to date.

A great deal of the fear of exposure to HIV among HCWs has arisen because they accept the notion that more education and training is required within their field, particularly around issues relating to values and attitudes, as well as protection and that they often feel powerless to contribute in the policy-making processes which directly impact on their work.

> McKinnon *et al.*'s study showed an increase in knowledge amongst hospital workers following distribution of an eight page booklet of HIV/AIDS information, and lectures from nurses experienced in AIDS care increased their knowledge base.
>
> (McKinnon *et al.*, 1990)

Negative attitudes occasionally persist in caring for patients with AIDS (PWA). Forrester and Murphy (1992) note that this often relates to the isolation of PWA, beliefs concerning testing patients' antibody status and whether they deserve to be cared for. Allied to this, there appears to be no relationship between HCW's level of education and their knowledge of HIV/AIDS (Barrick 1988), although an overwhelming conclusion was drawn that those with the shortest period of education, care assistants, demonstrated the most restrictive attitudes.

Younger HCWs also demonstrate greater fear for their personal health and that of their family, particularly if they have young children. However, it is identified by researchers that older HCWs may provide less 'hands on' care as they are usually in a supervisory or managerial role (Lester and Beard, 1988; Storosum *et al.*, 1991).

Concern has also been demonstrated by spouses and partners of those caring for patients with AIDS.

Klonoff and Ewers (1990) found that the two situations deemed most stressful by nurses, were having either friends or family know that they care for AIDS patients and having young children. Gallop *et al.* (1991) found that having children was one of the variables associated with extreme fear of contagion. One study of emergency personnel found that their families and friends were concerned that they might become infected as a direct result of their work.

According to further studies, prejudice still exists among care workers in their attitudes towards homosexual men. In their study of doctors in Canada, France and the USA, Shapiro *et al.* (1992) found that Americans were most likely to want to avoid care of homosexual men, PWA and injecting intravenous drug users. The French were least likely. The authors suggest that prevailing social norms may be important in relating these attitudes, although the studies were relatively small scale, and thus may not give an entirely accurate picture.

Attitudes towards women and sexuality have also been problematic among nurses. Webb (1988) found that nurses tended to steer the majority of discussions with their female clients towards the traditional route ignoring the fact that they may have other beliefs or values. Knowledge was also regarded as poor on tests given to nurses about sexual attitudes and knowledge. Webb also found that advisory situations were often vague, inaccurate or at times misleading.

Healthcare workers also demonstrate concerns that policies are being made by individuals who are not directly involved in the risks of caring for patients with HIV or AIDS. HCWs, particularly surgeons, feel that it would be beneficial for them to participate in policy decisions and protection measures as they are for the most part at increased risk. Their reasons are as follows.

1. **Lengthy procedures** – some studies have shown that the risk of occupational injury increases in lengthy operations or in procedures where HCWs are tired.

2. **Experience/inexperience** – the experience of a surgeon can often mean that others in the team are heavily reliant on that individual. This can lead to complacency on both sides.
3. **Emergency procedures** – studies have indicated that accidental perforation of latex gloves was increased threefold for procedures performed on patients who were unco-operative as a result of the influence of alcohol; this was sixfold if the patients were confused or unco-operative.
4. **Visibility** – Although extremely rare, there may be occasions where visibility is poor leading to accidents.

THE FOOD INDUSTRY

The likelihood of injury from the use of knives and other sharp implements means that there is at least a theoretical risk of individuals coming into contact with infected blood in the food preparation industry.

There are some companies who would not knowingly employ an individual who was HIV positive. The handling of knives and other potentially hazardous implements appear to increase the concerns within the industry.

EMPLOYMENT, LAW AND HIV

THE EMPLOYEE

Employees who are deemed HIV positive or who have AIDS have the same rights and obligations as any other employee. There are, however, specific points of which the employee should be aware.

- There is no obligation to inform the employer of the condition, although nursing and medical personnel are expected to follow their professional codes of conduct.
- Should the employee choose to disclose the information, it should be treated as a confidential matter. On occasions the employee may request that their supervisor or line manager be informed.
- The employee is likely to be advised to consult the company's occupational health/medical department; again this would be in confidence.
- Advice should be offered to take steps to minimize any risk of transmission of HIV to other employees. In most cases, it is likely to be the HIV positive employee who would be at greatest risk of contracting infections from others.
- It must be noted in the employee handbook or in the company's

policy document that unauthorized disclosure of a person's HIV status by any employee should be treated as a disciplinary matter.

- The employee should be offered the opportunity to receive counselling either externally or on the company's premises.
- If the employee later develops AIDS, then they should be treated as an employee who has a long-term illness. The employee may eventually need to be retired on ill-health grounds and would be entitled to any sickness benefit or pension accrued while in service.

ADDITIONAL MEASURES

One particular problem that employees who are antibody positive are likely to face is that others may refuse to work alongside them. Some companies, particularly in the public sector, have considered this situation. Some indicate that they are more likely to redeploy the individual(s) concerned in the event of succumbing to pressure from fellow employees, although it is emphasized that in the initial stages the education of the workforce would be a more tactful approach. The invocation of disciplinary procedures against the offender(s) would also be regarded as necessary.

> A sales representative at Digital Equipment Corporation, a large US computer manufacturer, had been aware for a number of years that he was HIV positive before he revealed his condition to colleagues at work. He was afraid of the potential responses of his colleagues, and wary of his job security, and well-being. Fortunately his employers were one of the corporate pioneers in instituting an education and advisory programme on HIV and AIDS. The programme was introduced in 1987 and included seminars to explain the disease, its progress and prevalence.
>
> By 1991 the employee, confident that the myths and fears surrounding HIV and AIDS had sufficiently abated, disclosed his condition to his colleagues. His departmental managers supported and assisted him, restructuring his job to take account of his need for medical treatment.
>
> (*Financial Times*, 20 January 1993)

Workplace colleagues may be offered advice on the risk and control of infection and how to cope in a humane and sensitive way with an HIV positive colleague. This, however, should only be carried out with the affected employee's consent. In addition to this, the employee's performance may be monitored in order to identify any special needs as they arise. The employee must also take care in the event of having additional responsibilities at work. For example, if the employee is a company first-aider, he or she may wish to withdraw from those duties. In the UK there is specific guidance outlined by the government to protect employee and employer alike. The guidelines state that to dismiss an employee purely on the basis of contracting HIV could be held to be unfair under the legislation. It is also the case if employers dismiss an employee affected by HIV as a direct result of pressure from other employees. Pressure to dismiss an employee backed by a strike or threat of a strike should be ignored as stated in the Employment Protection (Consolidation) Act 1978, Section 63.

Employers are also advised to investigate all complaints to ascertain the cause of any difficulties and consider ways of dealing with the problem. Employees who are discovered to harass or refuse to work with people affected by HIV may well themselves be in breach of contract and face disciplinary proceedings as a result.

During 1987 a worker was dismissed as a result of fellow employees gaining knowledge that he was homosexual and the fear that he might have AIDS. The industrial tribunal found the decision to be a fair one. Later, however, the man won his case claiming through the Employment Appeals Tribunal (EAT) who advised that the employer should pay compensation. The EAT decision was based on the fact that the company had not held a proper inquiry.

Employers should be aware that although no case exists at the time of writing, there is the increasing likelihood that male employees could take a case of HIV discrimination on the grounds of indirect sex discrimination, as there are currently in the UK more cases of men who are HIV positive than women. If a case were to be taken on the grounds of sex discrimination, no qualifying period of employment would be required. Currently it takes two years to present a case of unfair dismissal.

With regard to those employed in healthcare, the General Medical Council has ruled that doctors and health workers with HIV infection are ethically obliged to seek professional advice concerning the need to limit or modify their practice to protect their patients.

Employers who are aware of an employee's condition are ethically bound to respect the confidence and find suitable alternative employment at an equivalent level for the employee if there is deemed to be a real risk.

Currently, there are no professional or ethical grounds for refusing care to people with HIV. Any refusal to provide appropriate care and treatment can be a violation of professional standards of conduct or conditions of employment.

Attitudes to healthcare workers who become infected by HIV from occupational exposure are important in contributing to their perceptions concerning acceptability of the risk of infection. This group of workers need to be reassured that if they are infected in the workplace it will be accepted as a work-related disease and thus compensable. Wicher (1993) sites a case of a pharmacist being refused employment by a medical centre once his HIV status was disclosed. The practitioner was not taking part in any invasive treatment and so was not a risk to anyone. Cases such as this should be reviewed to prevent the likelihood of discrimination, and to protect the healthcare worker from experiencing any further stigma as a result of a lack of awareness on the part of employers.

THE TESTING DEBATE

To date, there is no law to prevent employers from requesting that a potential employee take an HIV antibody test as part of a company medical. The employer, however, must gain the employee's consent before the test is carried out. Doctors who conduct such a test without the individual's consent are committing a criminal offence and could be charged with assault and battery.

The issue of testing has, in certain countries, come increasingly into the public domain, since the issue of HIV and AIDs became dominant in the 1980s. The number of tests in Australia, for example, has been rising since records first became available. In 1987, the number of tests conducted in Public Health Laboratories in Australia totalled 354 348; in 1992 this had risen to some 814 127 tests. The majority, 366 387, were conducted in New South Wales. A similar trend has occurred in Sweden where almost 10% of the population has been tested for HIV. The increases have been for the most part attributed to education programmes advocating that if people are anxious about the risk of exposure then they should take the test.

Employers have no right to the test results. It is only the individual who may give consent to relay the result without undue pressure. The employer can, nevertheless, refuse to offer employment if the potential employee refuses either to be tested or to disclose the results of the test. The employer may also refuse to employ the individual found to be HIV positive if the type of work to be undertaken puts the antibody positive individual at risk.

As mentioned earlier, although as yet not tested in law, the employer should be aware that the potential employee can claim indirect sexual discrimination (if male) purely because they are deemed to be HIV positive.

For the most part, employers regard compulsory testing of employees as unreliable, expensive and rather time-consuming. The World Health Organization, the International Labour Organization, the Council for Europe and the British government are among those who have given a clear indication that there are no grounds for testing individuals for HIV either before or during employment. It would be beneficial to have a law preventing this practice as sadly some unscrupulous employers may pursue this discriminatory practice.

The US Centres for Disease Control (CDC) launched a 'Business Responds to AIDS' programme in 1992. The programme received support from major companies such as Levi-Strauss, Polaroid, Federated-Allied, Coca Cola and MGM/UA Communications. The programme involved the preparation and distribution of a manager's and supervisor's kit providing step-by-step guidance to developing a workplace policy and education programme. Counselling, resource and insurance information was also included. Some rather important advice was also provided for the employer:

- companies were discouraged from refusing to hire an applicant with HIV or AIDS because of the likelihood of a rise in insurance costs;
- companies were strongly advised not to ask potential employees to take a test to detect whether they are antibody positive;
- employers were also made aware that they cannot dismiss employees with HIV/AIDS if they are able to perform the essential functions of a job given reasonable accommodation to their disability.

The need for confidentiality is imperative. Employers are not legally obliged to disclose that an employee is infected with HIV unless the employee consents. The only exception to this is if it could be claimed that it was in the public interest for others to know. For example, this is common practice in the dental, medical and health-related professions.

Healthcare workers are generally against testing for the following reasons.

- Testing is impractical and expensive in terms of a considerable population working within the service and the relatively minimal risk of contracting HIV as a direct result of occupational exposure.

- They regard testing as problematic as it can lead to a false sense of security among staff.
- Having knowledge of those who are deemed antibody positive does not comply with the rules of confidentiality. This knowledge does not reduce the risk of exposure. Allied to this is the additional risk of discrimination that the affected member is likely to experience either directly or indirectly.
- Healthcare workers would have to be tested at least four times during a year. The estimated costs of testing all hospital doctors and clinical staff in the UK was in the region of £7.6 million in 1994, rising to £8 million in 1995. If midwives, nurses and dentists were included in the programme, the cost would be somewhat alarming at £65 million. This does not even account for ancillary employees such as hospital porters who deal daily with disposal of sharp instruments, needles and intravenous drug cannulae.

The very low risk of transmission from healthcare worker to patient and the protection afforded by implementing the 1993 guidance from the Department of Health diminishes the need for routine testing. Guidance also exists for education, environmental, public and funeral services where specific occupational groups may be at risk of becoming infected.

One cautionary note. Employers who disclose information relating to an employee's seropositive status must be wary of the serious consequences they are taking on. Such a disclosure can be detrimental to the HIV antibody positive employee and his or her family. The Data Protection Act 1984 (UK) holds in this case. Laws relating to confidentiality and HIV are further strengthened in Health Authorities by the National Health Service (Venereal Diseases) Regulations Section 1, 9, 1974. Additional protection is underlined in the Health and Safety at Work Act 1974, Control of Substances Hazardous to Health (COSHH) Regulations 1988 and 1994, and the Management of Health and Safety at Work Regulations 1992.

There are occasions when the need for testing may be beyond the control of the organization. One possibility is where an employee requires certification of sero negativity to qualify for an overseas visa. The second may be for insurance purposes. For the most part, however, few organizations envisage any circumstances in which testing might be introduced.

GUIDANCE FOR HEALTHCARE WORKERS

It is the responsibility of health authorities to formulate and provide specific detailed guidelines for all healthcare workers within their locality. For the most part, guidance will need to take into account issues relating to risk, compliance and safe practice.

In the consideration of risk, employers are required to assess the likelihood of potential occupational exposure. As mentioned, many health workers are in contact with sharp instruments and needles, blood and blood products on a regular basis. Adherence to the infection control procedures laid down by the health authority will assist in the reduction of risk of exposure. The Management of Health and Safety at Work Legislation 1992 also provides guidance in employee protection.

Compliance with regulations can be at times difficult and is at the time of writing considered to be imperfect. A study of British anaesthetists revealed that despite the existence of guidelines, up to 16% admitted not having worn protective gloves when carrying out certain procedures and at least a third of these continued to resheath needles. The researchers found that a number of issues were raised as to the likelihood of non-compliance. Firstly, there is often no time in emergency situations to don gloves or other protective equipment. This may well be the case if a casualty arrives unannounced in the accident and emergency department of a hospital who is bleeding profusely and requiring urgent treatment to prevent the condition becoming worse. It is equally the case if a casualty requires resuscitation.

The second reason for perceived non-compliance involves the low perception of risk. Healthcare workers are at greater risk of contracting hepatitis B infection than they are for HIV. This is established knowledge within the professions and hence the likelihood of mild complacency. Filtering of change is also a slow process within any organization and work situation, therefore the need exists for monitoring of the situation and regular updating of the knowledge of the workforce. This may assist in reinforcing the message without being too dogmatic.

The final reason relates to the interference with skill. Some dentists, for example, find it difficult to perform intricate tasks when wearing latex gloves. They also complain that they are uncomfortable to wear as they are so close fitting.

The development of safe practice is vital in minimizing the risk of transmission of HIV infection. It is wise for organizations to review their practice on a regular basis and to highlight any measures that are considered to be potentially hazardous or considered generally unsafe. The review and modification of infection control procedures will inevitably highlight the need for training and education.

POLICY AND THE WORKPLACE

The advent of the HIV/AIDS phenomenon during the 1980s saw a number of organizations developing policies to safeguard the employee and to give guidance to those working directly with those who were affected. For the most part early intervention came from the public services and large organizations. Indeed, the HSIB Survey (1991) indicated that in the UK, 94% of local authorities and 76% of health authorities had developed and were operating policies. The figure for private-sector companies was 51%.

Many organizations have had to consider their obligations to their employees should an employee become HIV positive. As a consequence, the development of a series of guidelines has been established. Many have been incorporated into existing policy relating to hygiene, health and infection control. The AIDS Control Act 1987 has also given employers incentives to attach the same level of importance to the issue of HIV and AIDS as they do to anti-discriminatory practice and equal opportunities in the workplace.

Details common to the majority of policies concerning HIV and AIDS include:

- the provision of basic information, in particular, that concerning the efficacy of the various methods of transmission of the virus, including a reassurance that the chance of transmission in the workplace is minimal;
- the need for confidentiality and assurance that the employee will be protected and that no testing will be carried out prior to or during employment under the auspices of the employer;
- a clear statement that potential employees who are antibody positive will be treated as any other job applicant; also, when they are employed they can be assured that their antibody status will have no effect on their continued employment within the organization; this includes promotion and training opportunities;
- a firm commitment that people with AIDS will be treated in the same way as any other person with a non-contagious debilitating serious condition. Thus the medical criteria at recruitment and employment will be whether the person is fit at the time of assessment and capable of satisfying the requirements of the job. An employee failing to meet these criteria will be entitled to the same sick leave, benefits and additional arrangements as any other seriously ill employee.

The overall responsibility for the operation of an AIDS policy lies with the personnel manager, although in many organizations as noted in the HSIB Survey (1991), the responsibility was shared. In some organizations, everyday implementation was the responsibility of line managers, although in health authorities departmental heads and officers shared it reflecting the wide scope of the policies relating to HIV and AIDS within the organizations. The majority of decisions taken in the structuring and formulation of policy were made jointly between management and union representatives within organizations.

KEY POLICY ISSUES

- Consider the prevention of discrimination at recruitment.
- Employees or potential employees should not be required to reveal their antibody status.
- Testing for antibody status should be left to the individual and not made a condition of employment.
- Any employee who has disclosed HIV status should receive treatment and benefits as any other employee.
- Special leave considerations are to be taken account of when caring for a partner or friend with AIDS.
- Any form of harassment should be regarded as a matter for disciplinary action.
- No employee should be dismissed on the grounds of their sero-positive status.
- Any redeployment of an employee who is HIV positive should be carried out with the employee's consent.
- Retraining should be offered as a matter of course where deemed to be necessary.
- When the employee's condition becomes impaired as a result of HIV-related illness, the same arrangements as are made for a person with a long-term illness should be employed.
- All employees should be given guidance relating to standard infection control procedures, as well as appropriate training and monitoring of the situation.
- Appropriate infection control equipment should be available, for example, paper towels, disinfectants, disposable gloves and aprons.
- In all healthcare settings, guidance is to be provided on the prevention of needlestick injuries.

EDUCATION AND TRAINING

The need for education and training of the workforce relates mainly to individual protection from the risk of contracting HIV infection, showing due care for others and the reduction of stigma surrounding the condition of HIV and AIDS. There is also the issue of confronting attitudes to sex and sexuality.

The medical profession appears to be taking measures to improve the situation relating to the issues of sexuality. One study found that up to 70% of physicians who graduated after 1981 said they had received formal training in human sexuality whereas only 53% of pre-1981

Lee Smith, former president of Levi Strauss' international operations and chairman of the Washington based National Leadership Coalition on AIDS, an educational organization focusing on business, recommends that given staff turnover and transfers, AIDS awareness programmes run continuously and be updated with new information.

. . . He warns that AIDS education is not an easy sell. There are terrible barriers around the disease. He noted that 1 in 25 people in Levi Strauss' San Francisco workforce was HIV positive. He intimated that managers, the majority of whom are heterosexual men and some of whom are homophobic, deny it is a problem and believe it will not affect them.

(*Financial Times,* January 1993)

graduates had received this type of training. However, it appears that more physicians were fairly comfortable about screening patients for substance abuse (77%) than they were with screening those with sexual concerns (54%). Education and training continues to be a developing process in the profession.

The New England Consortium for AIDS Education was set up by mine corporations to develop AIDS education materials. This spares individual businesses the costs of developing their own. The Consortium delivers a strong message to the businesses dealt with, that is to say, confronting issues relating to AIDS and the importance for businesses in dealing with problems that may arise.

Many organizations today produce information and advice to assist employers in giving and getting the correct messages across to their managers and employees. It is for the line managers to seek assistance from these agencies as current practice and the consistent updating of information will enable the workforce to become more enlightened in their approaches to the subject.

The Expert Advisory Group on AIDS formulated revised guidelines with the Department of Health for healthcare workers. The guidance incorporates statements relating to ethical considerations provided by a professional regulatory body. The Health Education Authority has also continued to produce materials to clarify information concerning HIV transmission and other sexually transmitted diseases, use of condoms and the negotiation of safer sex. Specific information is also provided relating to the reduction in transmission in certain occupations.

Employees in the UK, as with the general population, have access to the National AIDS Helpline. The new service began in 1995 providing assistance and answering queries relating to the conditions of HIV and AIDS. The service also provides guidance to those in need of specific services.

The Overseas Development Administration (ODA) along with the Medical Research Council are currently developing, improving and updating strategies for the prevention of HIV infection and the provision of cost-effective care. The project also includes research into vaccine development and is supported by 42 countries, as announced at the Paris AIDS Summit in 1993.

Any training and development strategy requires monitoring and evaluation for its effectiveness in getting the appropriate messages across and to note any cultural change within the organization. To date, there appear to be no studies which evaluate the effects of educational interventions for periods longer than three months. It is important to consider whether or not any enlightenment has had a sustaining presence within the work environment. Also the actual changes in behaviour need to be monitored over a considerable period. These will vary among occupational groups and range from consideration for others, reduction/increase in discriminatory behaviour and practice, use of protective clothing and open discussion relating to sexuality, HIV and AIDS.

The National Leadership Coalition on AIDS was founded in 1987 by a consortium of representatives from business, national health-care organizations, gay and religious groups and labour organizations. Belief in collaboration, co-operation and sharing of resources in society is their main aim. Their chief functions include the organizing of conferences, distribution of education materials and support of local businesses.

Examples of good practice appear to stem from proactive occupational health departments. British Telecom utilizes its network of occupational health nurses to maintain the nation's ongoing HIV/AIDS education campaign. These activities include the reinforcement of the principles outlined in policy, combined with practical advice on reducing the risk of transmission of the virus at the workplace and socially. The company also publishes a newsletter which provides news concerning the latest developments for employee protection from the company's chief medical officer.

Health authorities frequently join forces with businesses and colleges to educate employees. They tend to also employ the services of AIDS action groups and training officers to deploy useful knowledge and skills that are likely to be of benefit to employees.

Education and training in the workplace will include:

- consideration of national guidance and implications of company policy;
- addressing the issues surrounding sexuality and any false attitudes, myths, values and beliefs that exist relating to HIV and AIDS;
- addressing issues surrounding sensitivity and potential stigma;
- protection and rights of the employee;
- where to obtain further information and sources of assistance relating to counselling.

SUMMARY

HIV and AIDS is a condition that continues to be addressed by enlightened health authorities and companies. All employees are for the most part supplied with information that is readily available within the public domain such as modes of transmission and means of protection to reduce the risk of exposure. It is increasingly the case that employers need to introduce policies to address potential issues such as testing, protecting the affected employee and colleagues, and steps to take when the employee's condition becomes increasingly impaired.

SUGGESTED FURTHER READING

Advisory Committee on Dangerous Pathogens HIV (1990) *The Causative Agent of AIDS and Related Conditions*, Department of Health.

Department of Health (1995) *HIV/AIDS and Health Promotion: An Emerging Strategy*, Department of Health.

Expert Advisory Group on AIDS Guidance for Clinical Health Care Workers (1990) *Protection Against Infection With HIV and Hepatitis Viruses*, HMSO.

Forrester, D. and Murphy, P. (1992) Nurses' attitudes towards patients with AIDS and AIDS-related risk factors, *Journal of Advanced Nursing*, **17**, 1260.

HIV Infection and AIDS: Towards a Corporate Response (1993) Local Authority Associations Officer Working Groups on AIDS, Spring.

National AIDS Manual, National Aids Manual Publications Ltd, Eurolink Centre, London.

Recommendations of the Expert Advisory Groups On AIDS (1994) *AIDS/HIV Infected Health Care Workers: Guidance on the Management of Infected Health Care Workers*, March.

Wilson, P. (1992) *HIV and AIDS in the Workplace: An Examination of Cases of Discrimination*, London, National AIDS Trust.

REFERENCES

Barrick, B. (1988) The willingness of personnel to care for patients with Acquired Immune Deficiency Syndrome, *Journal of Professional Nursing*, **4**, 366.

Breault, A. and Polifroni, E. (1992) Caring for people with AIDS: nurses' attitudes and feelings, *Journal of Advanced Nursing*, **17**, 21.

Currey, C., Johnson, M. and Ogden, B. (1990) Willingness of health professions' students to treat patients with AIDS, *Academic Medicine*, **65**, 472.

Feldman, T., Bell, R., Stephenson, J. and Purifoy, F. (1990) Attitudes of medical school faculty and students toward Acquired Immune Deficiency Syndrome, *Academic Medicine*, **65**, 464.

Financial Times, 20 January 1993.

Financial Times, 15 February 1993.

Forrester, D. and Murphy, P. (1992) Nurses' attitudes towards patients with AIDS and AIDS-related risk factors, *Journal of Advanced Nursing*, **17**, 1260.

Gallop, R., Taerk, G., Lancee, W., Coates, R., Fanning, M. and Keatings, M. (1991) Nurses and AIDS, *Canadian Nurse*, **87**, 2931.

Health and Safety Information Bulletin (HSIB) (1991) AIDS at the workplace, **186**, June.

Klonoff, E. and Ewers, D. (1990) Care of AIDS patients as a source of stress of nursing staff, *AIDS Education and Prevention*, **2**, 338.

Lester, L. and Beard, B. (1988) Nursing students attitudes towards AIDS, *Journal Nurse Education*, **27**, 399.

McKinnon, M.D., Insall, C., Gooch, C.D. and Cockroft, A. (1990) Knowledge and attitude of healthcare workers about AIDS and HIV infection before and after distribution of an educational booklet, *Journal Social and Occupational Medicine*, **40**.

O'Donnell, L., O'Donnell, C., Pleck, J., Snarey, J. and Rose, R. (1987) Psychosocial responses of hospital workers to Acquired Immune Deficiency Syndrome (AIDS), *Journal of Applied Social Psychology*, **17**, 269.

Pomerance, L. and Sheilds, J. (1989) Perinatal nurses' knowledge and attitudes about AIDS, *Journal of Obstetrics, Gynaecology and Neonatal Nursing*, **18**, 363.

Scherer, Y., Haughey, B., Wu, Y.W. and Mathewson, M. (1992) AIDS: what are critical care nurses' concerns?, *Critical Care Nursing*, **23**.

Shapiro, M., Hayward, R., Guillemot, D. and Jayle, D. (1992) Residents' Experiences in, and Attitudes Toward the Care of Persons with AIDS in Canada, France and the United States. *Journal of the American Medical Association*, **268**, 510.

Storosum, J., Sno, H., Schalker, H., Krol, L., Swinkels, J., Nahuijs, M., Meijer, E. and Danner, S. (1991) Attitudes of health care workers towards AIDS at three Dutch hospitals, *AIDS*, **5**, 55.

Symser, M.S., Bryce, J.B. and Joseph, G. (1990) AIDS-related knowledge, attitude and precautionary behaviours among emergency medical professionals, *Public Health Report*, **105**, 496.

van Servellen, G., Lewis, C. and Leake, B. (1988) Nurses' responses to the AIDS crisis: implications for continuing education programs, *Journal of Continuing Education for Nurses*, **19**, 4.

van Wissen, K.A. and Sibers, R.W.L. (1993) Nurses' attitudes and concerns pertaining to HIV and AIDS, *Journal of Advanced Nursing*, **18**, 912.

Wallack, J.J. (1989) AIDS anxiety among health care professionals, *Hospital and Community Psychiatry*, **40**, 507.

Webb, C. (1988) A study of nurses' knowledge and attitudes about sexuality in health care, *International Journal Nursing Studies*, **25**, 235.

Wicher, C. (1993) AIDS and HIV: the dilemma of the health care worker, *AAOHN J.*, **41**, 282.

USEFUL ADDRESSES

National HIV Prevention Information Service, 82–86 Seymour Place, London W1H 5DB.

Family Connections, Adulona Lodge, 13 Tatam Road, Neasdon, London NW10 8HT.

National AIDS Trust, Room 1402, Euston Tower, 286 Euston Road, London NW1 3DN.

London Lighthouse, 111–117 Lancaster Road, London W11 1QT.

National AIDS Manual, Unit 407 Brixton Enterprise Centre, London SW9 8EJ.

Terrence Higgins Trust, 52–54 Grays Inn Road, London WC1X 8JU.

Healthy eating and the promotion of physical activity 7

Introduction • What is healthy eating? • Marketing healthy eating at work • What actions should organizations take? • Auditing healthy eating • The catering manager's role in promoting healthy eating • Developments in healthy eating • Monitoring and evaluating healthy eating initiatives: the role of the workplace health co-ordinator • Promoting physical activity at work • What is physical fitness? • Physical health at work: a case study • Summary • Suggested further reading • References • Useful addresses

INTRODUCTION

As long ago as the early 1970s, millions of American citizens read about a study conducted by researchers Belloc and Breslow (1972) who outlined the necessity of healthy eating and exercise for the maintenance of good health. They indicated that people should:

1. eat three regular meals each day and no eating between meals;
2. ensure that one of these meals was breakfast; and
3. exercise at least two or three times a week.

These rather simple but poignant facts were seen as a contrast to the reality of the busy executive or the construction worker. Indeed, they illustrated that many American workers grab a few minutes during the day for lunch and usually this consists of a sandwich or chips. While answering the phone, reading the newspaper or, today, working at the computer, the worker will balance a soft fizzy drink and french fries on a desk piled high with paperwork!

It is intended in this chapter to outline some of the developments in healthy eating and physical activity, as well as provide suggestions for managers to encourage, motivate and stimulate their workforce, but first, we need to look at what is meant by **healthy eating**.

WHAT IS HEALTHY EATING?

Interest in nutrition and nutrition policy during the mid-1980s arose in many countries as a result of the **Health For All** strategy outlined following the Alma Ata Declaration on Primary Health Care. One of the principles of the declaration listed under particular essential services was the 'promotion of food supply and proper nutrition' (Green, 1994).

The new concept of **healthy eating** arose, and was interpreted in Britain as an emphasis on the prevention of chronic disease, especially coronary heart disease. Nutrition education messages were based on **fat, fibre, sugar** and **salt**. Attempts were made to give useful information in the most practical way by using food-based messages, giving shopping and food preparation tips and introducing ideas for meals, as well as weekly menu planning and healthy choices when eating outside the home.

There appears to be a great deal of public awareness about diet and health; however, the practice of healthy eating is far from universal. The UK and US consumption of animal foods is high compared to that of some European countries, whose diet consists of more fresh fruit and vegetables and comparatively little meat. In the workplace, researchers have found that caterers see healthy eating as anything ranging from a fat-free diet, the avoidance of chips, to involving only the use of fresh produce. Maudsley's study (1989) found a frequent comment was the importance of a **balanced diet**; however, few catering managers were able to clarify this abstract term.

Allied to Maudsley's findings, it was noted that few catering managers appreciated the important role they had to play in facilitating any recommendations from health departments, government reports and the food industry. This would inevitably involve changes to the use of ingredients, food preparation techniques and menu planning to accommodate new proposals and varying attitudes of the workforce as a result of any change.

The provision of meals at work requires attention as it has considerable untapped potential in contributing to a healthy workforce. In Europe, the number of employees who take a cooked meal at work has risen steadily. In the UK, however, only about 20% of those establishments who are able to provide restaurant or canteen facilities do so. The case differs in Japan where employees in industry and commerce are offered subsidized meals as part of the company's health package. This is the case in at least 25% of workplaces. They encourage socializing at meal times in the *shokudo*, a communal dining area; 60% of the workforce generally make use of this.

THE IMPORTANCE OF DIET

The UK government's dietary targets for the year 2005 outlined in the *Health of the Nation* (DOH, 1992) strategy, represent a challenge not only for nutrition educators but also for those in agriculture, the food industry, advertising, food science, departments of public health and the customer. Much of the concern surrounds the population's total fat intake as a percentage of energy, which has been at least 40–42% over the last decade. The reduction in saturated fat from 17% to 11% of total energy represents a greater challenge. However, the most ambitious target is that of reducing the prevalence of obesity in 16–64-year-olds. Evidence to date indicates that overweight and obesity continue to increase to the extent that almost half of the adult population in the UK is currently above the healthy weight range for their height (Gregory *et al.*, 1990).

In 1940, the Ministry of Agriculture, Fisheries and Food's National Survey was set up to check the effectiveness of the government's wartime food and nutritional policies. Since then, it has continued to monitor diet in the UK to the end of rationing, when after years of compulsory restriction, Britain returned to a diet rich in meat, butter, eggs, sugar and white bread.

Convenience foods were introduced during the 1960s as patterns in employment changed and more women went out to work. This consequently led to higher consumptions of salt and sugar.

Changes in nutrition and food consumption in the early 1980s were prompted by the new interest in health and youth, where the introduction of health foods and products represented another change in diet. This was particularly the case in many American workplaces where new health promotion strategies were introduced.

The NACNE, COMA and BMA Reports (UK) published between 1983 and 1986, all recommended reducing the consumption of foods high in sugar, salt and fat; and increasing the intake of fresh fruit and vegetables and dietary fibre. Food consumption has, for the most part, fallen in line with these recommendations over the last few years, although consumption in fat continues to be problematic.

According to recent details of the Food Survey (Maff, 1993; as cited in Buss, 1995), the British diet continues to have more in common with northern Europe; meat and dairy products, bread, beer and potatoes remaining consistent in their proportions in the diet.

Healthy eating practices and policies continue to develop although the information provision can be inconsistent and ineffective. Indeed, the *Health of the Nation* document states that 'action to achieve health and nutrition targets involves the dissemination of information about healthy eating and encouraging and enabling changes in the populations diet'. Studies increasingly show that the public has a reasonable knowledge base about food and health. Across all socioeconomic groups, people are familiar with most of the broad healthy-eating messages (Sheiham *et al.*, 1990). They know that they ought to eat less fatty, fried foods, high fat processed and pre-prepared foods, cut down on sugary foods, and to eat more fruit and vegetables.

> . . . maximising the intake of bread, cereals, and potatoes. Over the last hundred years, consumption of cereals and potatoes has fallen as people eat more energy dense foods, heavy in fats and sugars. Low intakes of starchy foods are linked to the development of many diseases – coronary heart disease, strokes, cancers, diabetes. Cereals and potatoes are essential to prevent constipation which can lead to haemorrhoids and diverticular disease. Whole grain starchy foods are rich sources of essential fatty acids, B vitamins, minerals and trace elements.
>
> (WHO Recommendations)

The one main area of confusion and misinformation concerns the need to eat more fibre-rich starchy foods such as cereals, bread, pasta, potatoes and rice. These are carbohydrates and, until recently, no one would have been encouraged to eat them to stave off diseases such as cancer or heart disease. It is now apparent that foods rich in certain types of carbohydrate do exactly this (Cummings, 1995). There is increasing evidence that there are some health benefits linked to eating those carbohydrates that can be digested only in the large bowel rather than in other parts of the digestive system, where most carbohydrates are broken down. The benefits relate to the growth of specific bacteria that feast on the difficult to digest carbohydrates. The right combination of carbohydrates in the diet is likely, according to the research, to prevent the development of diabetes and bowel cancer.

There is, too, increasing evidence to suggest that the value of information giving through health education leaflets is questionable. Researchers have found this to be the case where leaflets are available for the consumer to pick up without any supporting advice from experts. Leaflets used in healthy-eating programmes have been shown to do little to enhance knowledge or effect dietary change (Murphy and Smith, 1993; Fowler, 1985; Nichols *et al.*, 1988).

A considerable gap exists between awareness and behaviour. Providing information about bodily functions, health risks and how to avoid them is not effective in bringing about specific behavioural changes to maintain optimal health. McKenna and Gowdy (1994) propose that the

> To identify the major areas of public knowledge and misconceptions regarding food and nutrition, a preliminary study was conducted. Representative samples of 1,800 consumers and 200 general practitioners and practice nurses was collected. Among the general public, it was found that 90% had heard of fibre, starchy foods and polyunsaturates but only 28% recognized saturates and 11% monounsaturates. When asked to identify nutrients in basic foods, 35% could only identify three sources of fibre and 75% recognized more than three foods containing starch; 83% failed to recognize foods which contain saturates, 92% foods which contain calcium, and 95% foods which supplied iron. It can be assumed therefore that although the public is generally aware of the 'buzz words' in nutrition, they cannot identify sources of nutrients very well. GPs and practice nurses were found to make good use of the wide range of expertise and information available in advising their clients.
>
> (Rudat, 1993)

key is to demonstrate that modification of diet is desirable and will have positive outcomes. This can be extremely problematic if outcomes are not immediately apparent. Persuading people to change long established eating patterns will require long-term strategies and combined efforts by food manufacturers, caterers, nutritionists, health experts and the individual consumer.

It is also unlikely that changes in diet and encouraging healthy eating will immediately occur without consideration of taste. Knowledge of the culture and social values of the client group is of considerable importance in promoting healthy eating. Dieticians, health educationists, and workplace catering managers are likely to be more successful in getting the message across if they have some understanding of the client's needs as opposed to delivering a blanket message which may be completely ignored.

Catering managers cannot assume that individuals will make use of the information with which they have been provided, nor can they assume that behaviour will be modified as a result of receiving health information. Catering managers will have the additional task of balancing other variables in food selection as well as the health benefits, such as financial constraints, motivation for eating, the influence of peer groups, early eating patterns and the amount of time available to consume food.

If an employee has expressed a need for further advice and support, then the manager should at least be able to provide a source of advice to which the employee can turn. Perhaps dietary counselling may assist in this process.

DIETARY COUNSELLING

Dietary counselling has increased over the last few years as an alternative option to slimming aids and reading information about diet. Information giving within dietary counselling is combined with motivational and behavioural components to assist the individual's approach to healthy eating.

Dietary counselling is used to describe a process which has many of the components of the counselling process. The client-centred approach consists of five stages which assist in clarifying or correcting misinformation.

PROCESS OF DIETARY COUNSELLING

1. **Screening/diagnosis** – this is based on the notion of risk or presence of actual disease and is usually the route through which clients are referred by the doctor to the dietitian or other appropriate member of the healthcare team for dietary advice.
2. **Assessment** – this stage involves the assessment of three different things: the client's readiness to change, their perceived barriers to change and their current diet. Establishing readiness to change is an important first stage of the initial consultation. If clients are not ready to change, they will need to be motivated. If they are ready, they may perceive barriers to change in which case different practitioner skills are required. Dietary assessment should only be undertaken once it is established that an individual is both ready and feels able to change.
3. **Negotiation and discussion** – discussion and negotiation about what is possible for the individual concerned is carried out, along with decision-making, so the programme becomes more realistic and meaningful.
4. **Goal-setting**– it is essential that the goals are decided upon by the client, without coercian or persuasion. The action plan should be agreed by both client and practitioner and there is evidence to suggest that compliance is enhanced if the client 'signs up' to the agreement, in the form of a contract.
5. **Monitoring progress** – permanent change in eating patterns is the main goal in dietary counselling. Relapse can occur, however, and shows that long-term behaviour changes cannot be expected from short-term nutrition education programmes. Monitoring progress is therefore an important aspect of long-term change.

Dietary alteration involves many complex changes not only in what an individual actually eats, but also and perhaps more importantly, in many other factors which affect food choice. The professionally led and persuasive approach has perhaps wrongly assumed that people will remember and act upon every piece of information given by the health professional, not least because the information is credible and legitimate. Dietary counselling takes the individual's needs into consideration and is useful to the employee who may decide that the group approach to weight reduction is not for them.

MARKETING HEALTHY EATING AT WORK

Corporate interest in nutrition over the last decade has arisen as a result of changes in policy and knowledge that many diseases contracted by the population are directly related to affluence and poor eating habits. Allied to this, the increased use of nutrition labelling as a marketing tool has prompted new controls regarding the presentation of nutritional information and claims. These developments have served to increase awareness of the importance of healthy eating which has, in turn, placed additional pressure on organizations to become involved in the production and marketing of food.

Carlson's work (1987) has demonstrated that managers lack the necessary background concerning healthy eating to instigate change. Those who were genuinely interested gained the majority of their information from professional magazines and the media.

The introduction of nutritional expertise has assisted in this process. Ruxton and Kirk (1993) surveyed a number of organizations employing corporate nutritionists. They found that 80% of those surveyed employed full-time nutritionists. Four companies indicated that they regularly employed the services of freelance nutritionists. A few public relations companies were also found to be employers of nutrition specialists.

The role of the corporate nutritionist is various.

1. *Attachment to medical companies* The reliance on nutritional knowledge and its application is expected in this area. Knowledge of health benefits of nutrition and health promotion techniques is necessary in order to provide appropriate information to the marketing and sales teams. This may involve consultations or input into copywriting.
2. *Attachment to food companies* The corporate nutritionist is employed to deal with customer queries, nutrional labelling, research and product development. In addition, the role includes the writing of leaflets as part of marketing information.
3. *Public relations* Nutritionists are heavily involved in the production of

documentation either for individual clients, as in the case of public relations companies, or during the promotion of generic foods such as milk and sugar. One of the aims of nutrionists working in this field is to brief and liaise with opinion formers such as dieticians, nurses, health visitors and doctors in order to reinforce a positive view of the product. In addition to appropriate documentation, this may involve the setting up of conferences or exhibitions or liaising with the media. They also keep abreast of current literature, conferences and liaise with associations such as the Nutrition Society, the British Dietetic Association and the Institute of Food Science and Technology.

The corporate nutritionist may also act as consultant to commerce, industry and education services with advice about techniques in getting the healthy-eating message across, as well as supporting catering managers in education and training.

WHAT ACTION SHOULD ORGANIZATIONS TAKE?

MANAGERS

Managers may appoint a responsible person in the organization to assess the current situation of food provision and menu preparation within the organization. Any needs should be identified and discussed with appropriate personnel. If training is required by the catering manager and relevant staff, this should encompass the medical and social benefits of healthy eating as well as connection to recent developments relating to diet and healthy eating. Any policy agreements and their relationship to employee health is to be considered. Caterers should be given the opportunity to discuss menus, nutritional content with nutritional experts and the time to pilot new dishes.

AUDITING HEALTHY EATING

This involves finding out what is currently available for employees at meal times, noting the most popular choices, listening to concerns and identifying any omissions on the menu. For examples of things to consider see page 135.

Once an audit has taken place, managers will need to consider whether they are prepared to work towards the promotion of healthy eating. They will need to make a clear commitment to achieving any targets set. These targets may be set within the organization, or proceed along a clearly defined programme such as those outlined by government or specialist organizations. Any nutrition activity should be integrated with other policies which promote health. Healthy-eating action proposals

AUDITING HEALTHY EATING CHOICES OF THE WORKFORCE

The following list could help to form the basis of future contracts with catering providers.

- What is available in the works restaurant?
- How is food presented?
- Is there adequate nutritional information for the workforce?
- What meal provisions are made for those working unsocial hours, breakfast meetings, evenings and weekends?
- Are healthy food options available for everyone including those from a variety of cultural/ethnic backgrounds?
- Are healthy options available in shops at the worksite?
- Are healthy foods options promoted at the point of sale to encourage people to try them?
- Are healthy eating principles reflected in the catering services providing food for the organization?
- Are customers given food in line with current dietary advice?

should be included in annual business plans and reviewed on a regular basis.

Practical implementation of healthy-eating programmes will involve a clear outline of the caterering manager's role.

THE CATERING MANAGER'S ROLE IN PROMOTING HEALTHY EATING

In Northern Ireland, the Healthy Eating Circle Award aims to encourage customers to change to healthier lifestyles by offering a wide variety of appropriate eating choices, optimal standards of hygiene and non-smoking areas. Catering managers in organizations where this incentive has been established have to balance both the complexities of surviving in business and the wider task of encouraging healthy-eating practices.

In adopting a policy of providing healthy meals to meet recommendations of government and health experts, catering managers are undoubtedly becoming involved in education and health promotion. Catering managers are responsible for planning menus and attracting customers, and if they are to fulfil and maintain this role, they should be educated and trained to achieve competence in health education.

Carlson (1987) points out that education and training of catering managers will involve changing their attitudes, minimizing stereotypical values and beliefs, as well as developing a culture where participating in health improvement and healthy eating becomes the norm. In his study of catering managers, Maudsley (1989) found that half the sample's understanding of healthy eating involved running promotional days or weeks. The impetus usually came from company managers or from 'pushy' dieticians for those involved in the health service. The extent of the promotions, when they occurred, varied from leaflets by the canteen till, to a complete change of menu. In one case, the healthy-eating initiative was seen as accommodating 'women on their health fads'.

One lesson that the catering managers did learn from their promotions was that introducing any change within the organization should be made gradually. This was particularly the case where the healthy-eating initiative was seen as a public relations exercise. The catering manager's staff team will also need persuading that healthy eating is good for the organization. After all, they are likely to be in the front line of service, so they will need to deliver positive messages to the workforce. Some of their creative ideas will also be of use to the catering manager in promotions and displays; thus it is vital that they are involved in training initiatives.

Catering managers, although acting as change merchants, will need to work within financial and personnel constraints. This is particularly the case in the public sector. Catering managers are for the most part given less autonomy in nominating suppliers and are bound by contracts which are out of their immediate control. This can make change difficult and reduce creativity. The value of nutrition will have to be balanced with conflicts in capital and quality.

Research carried out by the British Dietetic Association and Stanway amongst a sample of fifteen practising catering managers to establish perceptions and practice with respect to dietary guidance, found that the lack of suitable foods was the major obstacle in implementing dietary changes in their workplace restaurants. The researchers concluded that as at least eleven million people consume meals at work, catering managers are providing for a substantial sector of the population, therefore, they should be given a greater say in selecting suitable food suppliers.

(British Dietetic Association and Stanway, 1985)

Catering managers can support and encourage employees in making healthy choices by adopting current recommendations, principles and cooking methods relating to healthy eating. Nutrition experts suggest a series of measures for cooking and serving food with health in mind. The emphasis surrounding these measures are reductions in sugar, salt, fat, and the increased consumption of fibre, fruit and vegetables in the diet. Cooking methods such as steaming, grilling and the use of fresh produce are regarded as the best ways of introducing health into eating.

There is also emphasis in providing choice for the consumer. Individual employees should not be coerced into eating 'healthy' diets. Food is regarded as comfort as well as fuel; thus the health initiative should not be compromised with overkill – it puts employees off and is more likely to be met with rejection if the balance is not struck. In any menu change, catering managers should ensure there is variety and the calorific content clearly identified. The healthy option on the menu should be kept simple and be easily recognizable to the customer.

In promoting healthy eating in the workplace, the catering manager should always identify the means of:

- providing a good quality product;
- keeping prices keen but competitive;
- maintaining an attractive display of foods; and
- promoting healthy eating throughout the organization.

If these are taken into consideration then the chief executive is more likely to give support for the initiative, rather than regard it as 'yet another scheme'.

To encourage the take-up of healthy options, the following suggestions may be utilized.

- Plan healthy-eating promotions, e.g. raffle ticket with every healthy menu purchase, free meal or travel as a prize.
- Arrange sampling sessions with contract caterers, also perhaps in conjunction with local supermarkets or food companies.
- Place healthy choice meals at eye level to the customer and within easy reach.
- Encourage a warm, friendly, comfortable atmosphere to eat in. Some investment in training in customer care would be beneficial for the server, as well as providing them with first-hand information relating to the company health initiative.
- Make use of food spikes and menu boards to provide customers with information on dishes, prices and choices such as those high in fibre, and low in fat. If vending machines are utilized within the company, discuss with the suppliers the possibility of offering alternative selections of food that are healthy. For example, alongside the

chocolate bars and crisps, place fresh and dried fruit or even sunflower seeds.

- Devise sumptuous and attractive names for the dishes on the menu. Arrange the food in an attractive display. Provide salt and sugar substitutes in staff refectories and canteen; don't forget to offer decaffinated tea and coffee and fruit or herbal teas for those who request them.
- If possible, consider pricing the healthy option at slightly lower than other dishes offered. Any special dishes can be introduced occasionally as free tasting options. This may be done through the introduction of theme weeks, e.g. 'Thai week', 'Italian week', and so forth. Full use of promotional materials should be made, and if the company has a marketing department or access to the local press then the outcome will encourage other companies to participate in health promotion activities and events.
- For employees interested in weight reduction programmes, opportunities should be taken to introduce companies such as 'Weight Watchers'.

As mentioned earlier, one of the main tasks for the catering manager is planning the company's menu. This is by no means an easy task, as the catering for taste can be extremely various, as well as accounting for social and cultural needs. This is compounded by the additional task of adopting healthy-eating policies.

Introducing a new menu will take time and consideration of a series of factors. The catering manager will need to work alongside nutritionists, employee representatives and health advisers to ensure the correct balance is struck. Making use of available resources will also need to be taken into consideration in the interest of economy and making use of existing skills within the company. See box at top of page 139.

Catering managers have a key role to play in the promotion of healthy-eating initiatives within any organization. It is therefore imperative that any training and development initiatives should take account of developments already introduced within proactive organizations.

DEVELOPMENTS IN HEALTHY EATING

HEALTH AUTHORITIES

1. In the early 1980s Stepping Hill Hospital, Stockport, was found to have erratic food costs, high levels of plate waste, and general customer dissatisfaction. A working party was set up in 1981 to improve the catering. This included the implementation of a healthy food policy giving customer information and choice. Since then, standard recipes have been intoduced and continue to develop and change to suit

INTRODUCING A NEW MENU

Stage 1 Survey the kitchen staff and employees to ascertain their opinions relating to healthy eating and recent expert guidance on nutrition. Ask for their ideas and suggestions about foods served, atmosphere in the dining area, prices and waiting time.

Stage 2 Make use of current information and nutritional guidelines.

Stage 3 Take a realistic and objective look at the menu selection currently available in the organization, service area and profit margin. With consultation, modify recipes, cooking methods and service with health improvement in mind.

Stage 4 Address the issue of ingredients with nutrition experts and the company caterer. Suggested ingredients may include wholemeal bread, flour, and pasta, skimmed milk, lean cuts of meat, fish and poultry.

Stage 5 Keep a record of employee suggestions to inform menu and serving changes.

Stage 6 Ensure appropriate staff receive training, information and the skills they require to produce healthy food. Keep staff well informed about new developments and encourage them to contribute their own ideas and suggestions. Keep a suggestion box available to customers and staff and encourage them to be involved.

Stage 7 Promote healthy options to boost sales. Inform staff and customers about healthy options, give away leaflets and present the new menu sheet to customers. When launching the new menu, invite the local press, and health education representative to celebrate the launch.

Stage 8 Note any alteration in sales and consumption patterns of food, and keep a list of customer comments. Negotiate with wholesalers and manufacturers to obtain healthier products.

customer requirements. Plate waste was reduced as ingredients are being weighed and portions more accurately estimated.

The purchase of products has improved and commodities such as fat free milk, wholemeal flour and bread are included. An 'unacceptable products' list has also been devised to ensure customers are continually supplied with a quality product. Relations have also improved between catering staff and dieticians as well as establishing new ideas and techniques acquired from nurses and health education staff. Company surveys have indicated high levels of customer satisfaction and the majority of changes have all been introduced without an increase in costs.

2. NHS Hospital Catering is one of the largest catering organizations in the UK. It employs approximately 23 000 catering staff in 1700 units and provides 89 million meals to patients, NHS staff and visitors, costing in the region of £223 million.

NHS catering managers are fully versed in the strong links that exist between diet and health. To this end, they work in consultation with dieticians and others to ensure that they provide food which is both palatable and nutritious.

The caterers accept there is no room for complacency. They need to produce healthy, nutritious and appealing food in hygienic conditions, while at the same time responding to changes in the management of the health service.

> Bedfordshire's Healthy Workplace Project aims to evaluate the effectiveness of a workplace health promotion intervention to achieve a positive change in employee's lifestyles. This is being implemented on a policy, global and individual level through:
>
> - healthy-eating guidelines for food provision;
> - on site weight reduction/healthy-eating groups;
> - information activities focusing on a healthy-eating team.
>
> Alliances are being forged between Health Authority, Community Care Trusts and local business.

The Department of Health's Recipe File provides information for catering managers on healthy eating as well as a variety of suggestions for meals. These dishes include low fat, vegetarian, regional and cultural choices to stimulate creativity.

> Hertfordshire Health Authority UK, has an 'Enjoy Healthy Eating' initiative. A low-cost, low-maintenance scheme to introduce healthier foods and healthy lifestyles to students, schools and supermarkets in Hertfordshire has formed alliances facilitated by Hertfordshire's Health Promotion Department. Support has been given by one national supermarket chain providing a speaker for schools in addition to support from marketing boards.

3. The Health Education Authority – HEA (UK) has run a series of campaigns over recent years in order to promote healthy eating. They are:

- **'Less Fat Fortnight'** (LFF) June 1989
- **'Food For The Heart'** (FFTH) 1990, 1991 and
- **'Enjoy Healthy Eating'** (EHE) 1992.

The initiatives have utilized models traditionally used in the food industry. These include advertising, sampling, competitions and other public relations activities.

Less Fat Fortnight

This project ran for two weeks in June 1989 as part of the joint HEA and Department of Health's Look After Your Heart (LAYH) programme, and as a joint campaign with St Ivel 'Gold', whose interest was in promoting their product, a low-fat spread, as one simple way of introducing less fat in the diet. The HEA was keen to promote this, but at the same time to incorporate a balanced message concerning healthy eating. Such messages included simple dietary changes which could reduce the risk of coronary heart disease. As well as eating less fat, particularly the saturated type, the healthy-eating message also incorporated the increase in fibre-rich food consumption.

The HEA received support from 23 retailers, manufacturers and trade associations to assist in the promotion of the message. A further 134 health education units, dieticians and environmental health officers gave active support to the campaign.

THE WHO RECOMMENDS . . .

. . . Eating a minimum of 400 g of fruit and vegetables a day, including at least 30 g of pulses, nuts, and seeds. Vegetables and fruit are relatively low in energy (calories) but high in fibre, vitamins and minerals. Plant foods, particularly green and yellow vegetables and citrus fruits, seem to help prevent the development of cancers. These anti-cancer effects may be due to the vitamins C and E. A diet rich in plant foods is also associated with lower rates of coronary heart disease.

(WHO, 1990)

Food For The Heart

The overall objectives of the FFTH promotion were:

- raising awareness of the benefits of eating less fat and more fibre-rich foods;
- to assist the public in recognizing foods that are a valuable source of dietary fibre and to assist in dispelling the myth that fibre-rich foods are particularly fattening;
- to encourage members of the public to change their behaviour by purchasing lower fat foods and more fibre-rich foods in addition to encouraging alterations in food preparation and cooking methods;
- to encourage companies to make labelling and compositional changes to their products;
- to encourage active participation in the promotion by the food industry, workplaces and health education specialists, dieticians, and environmental health officers.

The specific messages promoted included:

1. eat less fat, especially saturated fat;
2. eat more fibre-rich foods;
3. simple dietary changes which can significantly reduce the risk of coronary heart disease.

The campaign's focus was the people of England, particularly those in the lower social classes, aged between 35 and 54 years, and men. Food retailers, trade associations and manufacturers were encouraged to participate by highlighting products which met specific nutritional criteria set by the HEA. Employers enrolled under the LAYH initiative were offered the opportunity to promote healthier eating among their employees. FFTH was supported by a national advertising campaign, mainly in the form of posters and advertisements in newspapers and magazines for women.

A comprehensive evaluation programme was undertaken of both the 1990 and 1991 campaigns. It comprised of a number of stages and employed a wide range of methodologies; LAYH public tracking study, shopper's exit survey, consumer discussion groups, employers' qualitative survey, workplace case studies, commercial partners' studies, health professionals workshop and evaluation studies.

Enjoy Healthy Eating

This campaign, launched in 1992, was introduced as part of a broader nutrition programme developed by the HEA. A greater emphasis was placed on the balanced diet and portraying healthy eating as enjoyable without making major alterations to the diet. The aims, objectives, messages and target audience were similar to FFTH, with the exception that the 'eat more fibre-rich foods' message was emphasized using the concept that meals should be based around such foods as bread, pasta, potatoes and rice. The promotional technique of a 'tasty challenge' was introduced to address one of the major barriers to change the perception that healthy food did not taste good. The taste challenge worked by providing the public with recipes and 'easy to do/hard to spot' tips in the HEA campaign literature. This was supported with tastings in supermarkets and local community events. They were also challenged to rethink their ideas on healthy eating, as well as to enter competitions which reinforced the message that meals should be based around fibre-rich starchy foods. The campaign was introduced in workplaces, supermarkets and hospitals.

COMPANY INITIATIVES

1. The move towards healthy eating began at ICI in 1982. Initially, it was decided that every restaurant would feature one starter, main course and sweet dish of a healthy-eating nature.

Following the introduction of the 'Heartbeat Award' scheme, in Wales (an initiative to combat the development of heart disease), the ICI restaurant at Pontypool won the award for two consecutive years. This led to the encouragement of other ICI catering managers in attempting to compete for the highly coveted award. The initiative had to overcome a few hurdles as they were unable to gain the support of the local environmental health departments and so could not be involved in the scheme. As a consequence, ICI started its own scheme. This gained considerable success at ICI's Watergarden restaurant, providing meals for around 3500 people.

2. Gardner Merchant is one of the largest contract catering companies in the United Kingdom. It currently serves over a million meals in organizations across the country. They have held healthy-eating initiatives since 1983, such as the 'Good For You' campaign, and publish their own healthy recipe booklets. The company has a commitment to promoting healthy eating and provides frequent alterations in menus to suit the customer and to accommodate any changes in food and healthy-eating policies.

MONITORING AND EVALUATING HEALTHY EATING INITIATIVES: THE ROLE OF THE WORKPLACE HEALTH CO-ORDINATOR

Any major initiative within an organization not only requires discussion, planning and execution, but also monitoring and evaluating the process and any alterations and developments that have occurred. This information will benefit the management team in terms of investment, training and provision of resources, as well as continue the momentum in developing further health initiatives, and finding the benefits to the organization and the individual employee.

Some monitoring and evaluation may involve the consideration of the catering managers' training, education of workplace staff, healthy-eating displays and sales and purchasing.

As indicated earlier, it is likely that training managers will need some training to implement healthy-eating policies. This is an important aspect of health education. Unless managers understand the reasons for the changes and how to introduce them, then any healthy-eating policy will be thwarted through lack of knowledge, apathy and incompetence.

Any long-term outcome evaluation of this training should involve improvement of sales or uptake of the healthier items on the works menu. In the short term, a questionnaire distributed to the catering managers at completion of formal training will provide some indication of immediate impressions and the value received from the training, and whether they have already started to put their new-found knowledge into practice.

Process evaluation will seek to answer some of the following questions.

1. Who was included in the training event? Why were these people chosen? Do they have a key function in the organization?
2. Where did the training take place? Was the venue conducive to learning?
3. What were the estimated and actual cost of training? Who paid for it?
4. How many catering managers were invited to attend?
5. What topics were included in the training and why?
6. What methods of training were used?
7. Which parts of the training event did the trainers feel were well executed and which aspects poorly received? What are the likely reasons for this? Draw on participants' reactions and comments.

By comparing the answers to these questions, discovery of the relevance and acceptability of such training will be found. The experience can be used to improve and develop further training.

As part of the implementation of a healthy food policy, many types of talks, lectures and presentations have to be made to many different groups of people such as company or local authority committees, a group of executives, a commissioning department, retailers or students. Before the talk, consideration will need to be given to the audience, the message and the way the message or request is to be communicated, as well as the desired outcome.

The outcome evaluation will in its simplest form be: did the audience get the message and did they make the desired changes?

Sometimes it may be appropriate to ask the group to complete a post-talk questionnaire. For other groups this method may be inappropriate.

Process evaluation should seek to address the following questions.

1. Why was the presentation made to the group?
2. Who was included in the group and why?
3. Why was the venue chosen?
4. What was the content of the presentation?
5. How did the audience react to the information, to the way in which it was presented and to the speaker?
6. What follow up is needed? By the group? By the presenter?

Displays may take the form of a selection of healthy foods being served in the company restaurant, or a display board encouraging lower fat consumption. One can assume that the need has been identified and the display has been designed to meet that need. Any outcome evaluation would seek to produce evidence of change in awareness of issues, knowledge, attitude or behaviour as a result of the display. The process evaluation may include an assessment of how any alterations in behaviour were achieved, for example:

1. Who constructed and organized the display and how long did it take them?
2. What was the cost of the display?
3. How was the display used? Was it unattended or were promoting staff present?
4. To what extent did the display attract attention? How many people looked at the display and did they just glance, or observe the detail and seek further information?
5. Did one part of the display attract more interest or cause more comment than another?

Continuing evaluation of this nature will produce the effect of ensuring that the best use is made of the display. This will inevitably assist in developing an understanding of how the changes in knowledge, attitudes or awareness measured in the outcome evaluation are achieved.

The catering department provides food for the workforce. There are many ways in which trends can be monitored, and sales and attitudes can be ascertained.

Monitoring food sales is one way of assessing the possible effects of a healthier eating promotion. For example, observation of sales in hospital staff restaurants before, during and after a promotion. The till roll can be used to provide information on sales. This will be possible if the company's accounting system is easily accessible and accurate records kept.

The catering manager will initially be interested in total sales. S/he will not welcome any promotion activity which reduces sales, but will be eager to build on and extend any promotion which increases their turnover and profit. This is equally the case for institutional caterers as for those in the private sector.

Monitoring the sale of specific food items and time of purchase will provide some indication of the food choices being made by the workforce, and whether these are changing. This detailed monitoring is only possible if the till system records which food items are sold and if the staff operating the till are fully aware of the codes to be used when registering specific items.

If meal times are staggered within the workplace, then the time of sale of a food item may be used in identifying differences in food purchasing by different sectors of the workforce. For instance, if the shopfloor workers lunch break is 12 noon until 1.00 pm, and white-collar workers is 1.00 until 2.00 pm, then a comparison can be made in terms of food preferences and uptake of the promotion messages and healthy food choices between the two groups.

Changes in purchasing may not necessarily be the result of healthier eating. Other factors may account for the change. These may include seasonal variations, daily variations in weather conditions, changes in staff, changes in shift systems, changes in the food offered by rival outlets, external advertising campaigns, alterations in the quality and attractiveness of meals, changes in pricing and general taste. Despite this, it will be beneficial to ascertain how changes in eating patterns have arisen even if one is not entirely sure how they occurred, after all, health initiatives arise outside the workplace as well as in it.

PROMOTING PHYSICAL ACTIVITY AT WORK

The Confederation of British Industry (CBI) UK survey in 1993 found that only a quarter (24%) of workplaces undertake initiatives to promote exercise. It has only been within the last decade that researchers have directly began to question the health benefits of exercise. Previous studies have concentrated on the mechanisms underlying adaptations to exercise or on coronary heart disease, and whether there is an association between physical activity and disease.

The Royal College of Physicians (1991) finally concluded in their report that

> there is substantial evidence that regular aerobic exercise such as walking, jogging, dancing or swimming is beneficial to general physical and psychological health.

The report reviews the evidence that regular exercise is effective in preventing coronary heart disease and osteoporosis, and concludes that exercise does have an important role in preventing these conditions. Physical exercise also has value in the management of other important conditions such as obesity, late onset diabetes mellitus and musculo-skeletal conditions.

If it is the case that 'vigorous' exercise is the important factor in improving heart health, then the proportion of those who take exercise of sufficient intensity appears to be very low indeed. Some researchers found that only about 9% of their sample of male civil servants performed exercise which met the definition of vigorous exercise, that is they participated in some activities which included walking and cycling. Most people who do take exercise usually do this in irregular bursts during their leisure time, or at work where, for the most part it may not be beneficial to health.

WHAT IS PHYSICAL FITNESS?

Physical fitness means different things to different people. Some definitions include:

- a person's ability to utilise the machinery of their body in sports and exercise;
- the ability to carry out daily tasks with vigour and alertness, without undue fatigue and with ample energy remaining to enjoy leisure time pursuits and to meet unusual situations and unforeseen emergencies.

The biological perspective refers to the body's running smoothly and efficiently. In recent years, there has been growing awareness of the benefits of maintaining a physically fit body. That is to say, improved health in reducing the risk of cardiovascular disease, improved resistance to infection, improved respiratory fitness. It may also assist in reducing the effects of asthma, poor posture and obesity.

Physical fitness, it has been suggested, can also contribute to:

- improved co-ordination and mental alertness;
- relaxation as it reduces the effects of stress;
- personal expression;
- increased longevity;
- social confidence.

PHYSICAL HEALTH AT WORK: A CASE STUDY

During the 1980s, Peter Conrad (1987) carried out a study into workplace health promotion activity in the Medical Technology Company, USA.

The company comprised employees involved in sales, administration, technical/scientific and various tiers of management. It started its health and fitness programme in 1983 as a beneficial initiative with no particular goal in mind.

The programme was open to all employees on a voluntary basis. Each employee paid $52 a year, and at least a third joined the scheme in its first year. It was administered by an outside vendor and staffed by a full-time health and fitness director. The vendor was a medically based organization that provided support, training and medical back-up for their on-site health promotion co-ordinators. The programme is housed largely in the health and fitness centre, which consists of locker rooms with shower facilities for men and women, two sauna suites, a 650 square foot exercise area and weight and exercise equipment room.

The programme attempts to offer a comprehensive approach to health promotion. Upon joining, each participant is given a medical evaluation that includes a computerized health risk appraisal, a series of blood tests for cholesterol, sugar content, blood gases and so forth. Tests also include measuring and recording blood pressure. This is followed by an individual counselling session. The classes are scheduled before and after work, and during the lunch hour, so are in the employee's own time.

Participants can sign up for as many classes as they wish. The range includes aerobics, stretching, jogging, nutrition, smoking cessation and weight control. While classes are run throughout the year, there are in addition monthly educational seminars relating to topics such as stress control, low back pain, diets and health foods. The programme director provides follow-up literature and individual counselling on a weekly basis.

The majority of participants within the programme identified their main aim as weight reduction and staying in shape. According to Conrad, there is little doubt that regular vigorous exercise itself had positive benefits. He notes that exercise does not only provide health benefits, it also sparks a more positive attitude to work, enhances the corporate image and increases competition in the international marketplace.

In conclusion, a useful agenda for action should include:

- provision of shower and changing facilities;
- provision of simple aerobic exercise equipment;

- making use of external organizations to form a comprehensive programme, e.g. local leisure clubs, LAYH specialists;
- provision of social facilities;
- employer's partial contribution to membership fee.

SUMMARY

Any healthy-eating initiative will require consideration of the needs of the workforce, suitable resources and provision of training for those who will be at the centre of change. Policy developments and new techniques based on sound evidence will also need to be incorporated into the catering manager's agenda. Monitoring and evaluation of various parts of the process of change will assist in continuing to inform those at the forefront of change.

Introduction of physical activity in the workplace will need support from senior management especially if a substantial financial layout in resources is required.

SUGGESTED FURTHER READING

British Dietetic Association and Stanway, A. (1985) *The Great British Diet*, Century.

Conrad, P. (1988) Health and fitness at work: a participant's perspective, *Social Science and Medicine*, **26**(5), Pergamon Journals Ltd, 545–50.

Hunt, P. (1995) Dietary counselling: theory into practice, *Journal of the Institute of Health Education*, **33**, 1.

Kemm, J.R. and Booth, D. (1992) *Promotion of Healthier Eating: How to Collect and Use Information for Planning, Monitoring, and Evaluation*, University of Birmingham, HMSO.

Morgan, E. (1989) The food industry in the year 2000, *Nutrition and Food Science*, March/April, MCB University Press.

Royal College of Physicians (1991) Medical aspects of exercise: benefits and risks, *Journal of the Royal College of Physicians*, London, 25, pp. 193–6.

Ruxton, C.H.S. and Kirk, T.R. (1993) What do nutritionists in industry do?, *Nutrition and Food Science*, **1**, Jan./Feb., MCB University Press, 22–3.

Taylor, D. (1992) *Human Physical Health*, Cambridge University Press.

Wallace O.P.G., Brennan P.J., Haines A.P. (1987) Are general practitioners doing enough to promote a healthy lifestyle? Findings of the Medical Research Council's General Practice Research Framework on Lifestyle and Health, *British Medical Journal*, **294**, 940–2.

REFERENCES

Belloc, N.B. and Breslow, L. (1972) Relationship of physical health status and health practices, *Preventive Medicine*, **1**, 409–21.

British Medical Association (1986) *Diet, Nutrition and Health*, A Report of the Board of Science.

Buss D.H (1995) How Healthy is the British Diet?, *Journal of the Institute of Health Education*, **33**(1).

Carlson, B.L. (1987) Promoting nutrition in the commercial food service establishment, a realistic approach, *Journal of the American Dietetic Association*, **187**(4).

Committee on Medical Aspects of Food Policy (1984) *Department of Health and Social Security: Diet and Cardiovascular Disease*, HMSO, London.

Confederation of British Industry Survey (1993) cited in Physical Activity: An Agenda for Action, *National Forum for Coronary Heart Disease Prevention*, August 1995, 31.

Conrad, P. (1987) Wellness in the workplace: potentials and pitfalls of worksite health promotion, *Millbank Memorial Fund Quarterly*, **65**, 255–75.

Cummings, J. (1995) A new look at dietary carbohydrates, *MRC News*, Autumn.

Darrall, J. (1992) The awareness and reactions of the food chain to healthy eating, *Nutrition and Food Science*, **23**(1), MCB University Press, 7–8.

Department of Health (1992) *Health of the Nation: The White Paper*, HMSO, London.

Fentem, P.H. (1992) Exercise in prevention of disease, *British Medical Bulletin*, **48**(3), July, Scientific Editor, J.B. King. Churchill-Livingstone, 477–716.

Fowler, G. (1985) Health education in general practice: the use of leaflets, *Health Education Journal*, **44**, 149–55.

Green, A. (1994) *An Introduction to Health Planning in Developing Countries*, 54, Oxford University Press, Oxford.

Gregory, J., Foster, K., Tyler, H. and Wiseman, M. (1990) *The Dietary and Nutritional Survey of British Adults*, HMSO, London.

Maudsley, C. (1989) Catering for health, *Nutrition and Food Science*, May, MCB Press, 2–3.

Murphy, S. and Smith, C. (1993) Crutches, confetti or useful tools? Professional views on and use of health education leaflets, *Health Education Research: Theory and Practice*, **8**(2), 205–15.

NACNE (1993) *Proposals for Nutritional Guidelines for Health Education in Britain*, Report to the National Advisory Committee on Nutrition, Health Education Council.

Nichols, S., Walter, W.E., Moolaway, M. and Hamilton-Smith, M.B. (1988) Evaluation of the effectiveness of a nutritional health education leaflet in changing public knowledge and attitude about eating and health, *Journal of Human Nutrition and Dietetics*, **1**, 223–38.

Royal College of Physicians (1991) Medical Aspects of Exercise: Benefits and Risks, *Journal of the Royal College of Physicians*, **25**, 193–6.

Rudat, K. (1993) Getting the message across, *Nutrition and Food Science*, **1**, Jan./Feb., MCB University Press, 35–7.

Ruxton, C.H.S. and Kirk, T.R. (1993) What Do Nutritionists in Industry Do? *Nutrition and Food Science*, **1**, Jan./Feb., MCB Press, 21–3.

Sheiham, A., Marmot, M., Taylor, B. and Brown, A. (1990) *British Social Attitudes Survey Report*, SPCR.

USEFUL ADDRESS

The British Nutrition Foundation, 15 Belgrave Square, London SW1X 8PG.
The Central Council of Physical Recreation, Francis House, Francis Street, London SW1P 1DE.
The Food Commission, 88 Old Street, London EC1V 9AR.
The Sports Council, 16 Upper Woburn Place, London WC1H 0OP.

Repetitive strain injury 8

Introduction: defining RSI • Predisposing factors to RSI and links with the workplace • Visual display unit operators • The meat packing industry • Prevention and workplace adjustments • Summary • Suggested further reading • References • Useful addresses • Notes

INTRODUCTION: DEFINING RSI

Bernardi Ramazzini (1713) was the first to detect what is known today in Britain as repetitive strain injury (RSI). The principle causes of this condition relate to psychological stress, working posture and repetitive motions. It is predominantly linked with the workplace.

The type of work performed over the years has led to an increasing recognition of the problems particularly relating to musculoskeletal activity. This includes machine pacing and the Taylorist fragmentation of tasks with its series of repetitive motions.

The condition of repetitive strain injury has not been an easy one to define as there are a number of musculoskeletal disorders associated with the neck, shoulders, arms and hands. This has led to controversy in the field of study. However, the most notable terms come from a variety of sources.

- occupational cervicobrachial disorder (OCD) – Japan
- upper limb syndrome
- shoulder arm syndrome
- overuse syndrome – Scandinavia
- cumulative trauma disorders (CTDs) – USA and Canada
- neck and upper limb disorders (NLDs)
- repetitive strain injury (RSI) – origins in Australia, later taken up by the UK

Neck and limb disorders (NLDs) were originally defined by Kuorinka and Viikari-Juntara in 1982, and divides the symptoms of the condition in working populations into five categories.

1. Temporary fatigue, stiffness or soreness of the muscles comparable to the muscle pain of athletes.
2. Long-lasting pain in the muscles, and tendons.
3. Symptoms relating to fibrosis of the muscles, that is, pain and swelling.
4. General disease with musculoskeletal symptoms, e.g. degenerative and rheumatoid arthritis.
5. Marked musculoskeletal pain in persons prone to depression.

Cumulative trauma disorders (CTDs) is the term used in the USA and Canada and refers to disorders of soft muscle tissue as a result of repeated exertions and movements of the body.

Occupational cervicobrachial disorder (OCD) was defined by a committee of the Japanese Association of Industrial Health in 1973 as the functional and/or organic disturbances resulting from neuromuscular fatigue due to carrying out jobs in a fixed position or with repetitive movement of the upper extremities. The Association also note that mental and environmental factors should not be ruled out.

The Australian term, repetitive strain injury (RSI) is described as a collective term for a range of conditions characterized by discomfort and persistent pain in the muscles, tendons and other soft tissues, with or without physical manifestations. RSI is also aggravated by repetitive movement, sustained or constrained postures and/or forceful movements. Psychological factors, as well as stress at the workplace, are also associated with the development of RSI.

The relatively recent associations with RSI broke through in the early 1980s in Australia where there was the announcement of an epidemic peaking around 1985 and 1986. The main workers affected included electronic keyboard operators and those in metalwork production industries.

The epidemic became pronounced as affected workers presented symptoms which differed from the well-known tenosynovitis. The workers appeared to be suffering from a chronic pain malady, encompassing muscle damage, psychological disturbance and neurological sensitization. Workers also experienced pain in the neck, upper limbs and shoulders. The victims of RSI lay the blame for their condition firmly on the work activities they perform.

Reporting of this condition under the term RSI, has been the practice in Britain since the early 1990s. It came to the forefront of the media when K. Tovey, an employee of the Inland Revenue, claimed and received damages in the region of £82 000 as a direct result of the work task she had to perform which led to the development of the condition.

Occupational medical researchers in Finland have classified RSI under three main disorder groups:

- occupational cervicobrachial syndrome;
- lateral elbow pain syndrome (tennis elbow);
- peritendinitis and tenosynovitis.

OCCUPATIONAL CERVICOBRACHIAL SYNDROME

This syndrome, originally discussed by Hult (1954), refers to generalized neck and shoulder pains which connect with the cervical group of bones of the spine.

Neck pain radiates into the shoulder region and down the arm. There is also reflective pain in the upper back muscles. The muscular pains in these conditions are numerous, hence difficulty in studying and classifying them as a separate group of disorders. Other common complaints include a deep diffuse muscular ache from the shoulder blades radiating out towards the level of the elbows. There is also tingling and numbness of the fingers, sometimes accompanied by episodes of dizziness, fainting or tinnitus.

Although not a definitive list, occupational cervicobrachial syndrome has been associated with teleprint operators, sewing machinists, microscopists and laboratory workers. It has also been found in VDU operators, cash register operators, typists, office workers and nursery teachers (Grieco, 1986).

LATERAL ELBOW SYNDROME

Overuse of the muscles of the wrist and fingers causes pain which radiates towards the lower arm and elbow. This is known as **lateral epicondylitis** or tennis elbow.

Tenderness and swelling arise as a result of tendon damage, inflammation of the ligaments and entrapment of the radial nerves.

The disorder results from repeated forceful movements of the wrist and forearm; that is, flexion and rotation. It appears to have been more pronounced among assembly line workers (during wrapping and wiring tasks) and among lumberjacks. The condition is also associated with tennis players, golfers, squash and badminton players.

PERITENDINITIS AND TENOSYNOVITIS

Peritendinitis is a condition arising from muscular overuse where the tendons and surrounding tissue within the body becoming inflamed.

This can, if left untreated, lead to oedema[1] and fibrosis[2] concentrated around the muscle-tendon junction.

Tenosynovitis is a disorder affecting the tissues of the hand, wrist and arm. It is slightly different to peritendinitis, in that it is manifested by inflammation of a tendon in its sheath and surrounding tissues. Sites often affected include the wrist, hand, fingers, forearm and elbow. Common signs and symptoms of this condition include:

- painful or aching fingers, wrists or lower arms – this may be a nagging pain similar to toothache;
- pain in the elbow worsening on movement;
- swelling of the affected area;
- loss of function and/or weakness of the grip of the hand;
- crepitus[3] over the affected area;
- numbness and/or tingling in the fingers, hands or wrists.

Tenosynovitis occurs as a result of rapid repetitive movements of the lower arm, wrist and fingers. It has been recognized among assembly line workers since the 1930s. It occurs in workers who are required to undertake tasks using manual force, or involved in awkward and unsuitable positions. This condition is common among meat and poultry workers, engineers and VDU operators (English *et al.*, 1989). The condition has also been detected in agricultural workers, e.g. cane cutters, net makers and hop pickers.

PREDISPOSING FACTORS TO RSI AND LINKS WITH THE WORKPLACE

Wallace and Buckle (1991) note that evidence linking occupational exposure and causal relationship of RSI as an illness presents some difficulties for researchers. For example, collection of exposure information is usually reliant upon the memories of respondents. This can lead to bias in recall and reporting.

In addition, they point out that simply recording the number of years' service in the same job is insufficient as one needs to have information relating to the whole work, sport and domestic activity history to gain accurate records of exposure to RSI. New technology may also be introduced in the workplace which can rapidly alter the long-term and cumulative effect on health, skewing any research results already obtained. Any results from research have to bear these factors in mind. However, the majority of findings do undoubtedly link the condition to the workplace.

The following factors have been associated with RSI.

1. **Posture** Poor posture is associated with poor and unsuitable work layouts. Employees consequently assume unnatural positions

involving reaching, bending, stretching or leaning. This results in significant pressure on the upper limbs and surrounding tissues. Laboratory and field studies of various occupations have sought to prove evidence of constrained postures and a variety of musculo-skeletal complaints. The range of findings includes pain in the hands, arm, shoulders and neck. Postural fixity has also been detected involving long-lasting static constriction of the muscles. This can lead to a reduction in blood flow within the muscles causing tiredness, pain and cramps (Hunting *et al.*, 1983 and Grandjean, 1984). Grieco (1986) has also noted the relationship between postures and cervico-brachial disorders. His findings indicate a high incidence in occupations involving VDU operations, and cash registration. The highest incidence he has found is among sewing machinists where 90% of 40 operators who had a short working cycle sat for at least six hours during their working day.

2. **Force** Some occupations require workers to exert considerable (manual) force using their hands or through their tools. The resultant pressure on limbs, joints and muscles is associated with poorly designed work areas, incorrectly sited machinery or type of tool unsuitable for the tasks to be performed because they are outdated. Maeda *et al.* (1980) found that line-assembly workers in a cigarette factory experienced cervicobrachial disorder as a result of heavy asymmetric loading of materials.
3. **Repetitive movements** Work involving speed and carried out in short cycles such as machine pacing in assembly production, can place the worker at increased risk of developing RSI.
4. **Particular hand movements** The use of tools, e.g. grippers, pliers, hammers necessitates forceful gripping, reaching, moving and twisting. Some work may also involve operating at awkward angles or feeling the force of excessive vibrations from electrical tools: both can lead to RSI.

Research to date has identified particular difficulties in occupations where RSI has become predominant.

VISUAL DISPLAY UNIT OPERATORS

The majority of studies relating to RSI has been in the use of visual display units (sometimes referred to as visual display terminals, VDT).

The management style and relative flexibility within the workplace apparently reduces ill health effects on employees. The opposite is, however, frequently found where VDU operators employing simplified tasks, according to Cakir *et al.* (1990), tend to suffer more fatigue, back and neck pain, as they are allowed fewer break times.

Smith *et al.* (1980) carried out a study of VDU operators at three work sites. It was noted that operators at one site had significantly more musculoskeletal, visual and other health complaints than operators at the other two sites. The jobs at this particular location involved inflexible work procedures with employees under constant pressure to perform and little operator control. This was the opposite to the other worksites whose employers allowed a degree of flexibility, autonomy, control and enormous job satisfaction. The main concerns that arose from these two sites was over career development and future prospects where more psychological difficulties such as irritability and anxiety were exhibited.

Elias *et al.* (1980) found in their study of bank clerks, that in employment data acquisition involved fragmented tasks, poor cognitive content and an extremely short working cycle led to RSI difficulties. The complaints noted involved neck, shoulder, back pain and notably poor job satisfaction. These elements also have an association with stress.

Dainoff (1982) identified in his reviews a relationship between health status and the type of work performed. Tasks involving data acquisition and entry operation yielded a higher incidence of musculo-skeletal difficulties among VDU operators than among those VDU workers performing programming, dialogue and editing tasks.

Nishiyama *et al.* (1984) saw the organization of working operations as a causative factor in aggravating posture and inducing neck tension and shoulder pain in three groups of newspaper workers. Of those studied, 437 VDU operators reported a significantly higher prevalence of shoulder and upper back pain, as well as pain in the fingers, than other keyboard operators. The way that workstations were set up, it was found, did not account for predominance in right- or left-hand use. VDU operators were involved in keying in information to their computers with the left hand and the right hand was used for the operation of cursor keys or the joystick.

Others in the same study complained of visual disturbances and long-waiting times to process information, the frequent need to seek advice from colleagues or to refer to manuals. This resulted in incomprehensible errors and repetitive operations. All reported increased bodily tension and irritability. Training, or the lack of it, was a major contributory factor to the development of RSI.

THE MEAT PACKING INDUSTRY

RSI has been found among a variety of workers in the meat and poultry industries (Roto and Kivi, 1984; Armstrong *et al.*, 1982; Buckle and Baty, 1986).

The shape, nature of the product and the tasks undertaken have contributed to problems relating to movement, posture and force.

Viikari-Juntura (1984) found that in slaughterhouses workers developed tension neck syndrome, tenosynovitis and peritenditis of the wrist and forearm.

In Canada, where reporting of RSI has increased since the mid-1980s, the National Work Injuries Statistics Program of Statistics indicate that the total number of injuries reported in the Canadian Meat Injury increased from 9731 in 1982 to 11 717 in 1986. This indicated a rise of 32% in the period. The highest rate of increase in RSI was noted in upper limb strains and inflammation being 58%.

The occupation of meat packing has a hazardous element to it, with butchers' work demanding hand motions with knives, and the risk of laceration, resulting in a tendency towards a high degree of muscle strain in the hands, arms and wrists. Novek *et al.* (1990) have indicated that various factories in the meat industry have undergone restructuring to extend the scope of managerial control, to reduce the value of labour and to submit the labour process to a regime of machine-paced repetitive work. Market uncertainty has also contributed to the increased risk of RSI, where fewer workers have had to take on more work and operate at a much quicker pace.

PREVENTION AND WORKPLACE ADJUSTMENTS

Prevention of RSI is not simple. Development of the condition occurs as a result of a variety of job tasks which occur, for the most part without deliberate intent. The keys to prevention have been found in adjustment of the work area, noting the use of tools and good posture. Legislation in the UK also impresses upon employers the need to take adequate precautions.

ADJUSTING THE WORK AREA

The work space that an individual works at should be the correct height for that person, and easy to reach without undue stretching, turning, reaching or twisting. Adjustments should be made for the individual worker if it is an area where the majority of work will be undertaken. Easily adjustable chairs, arm rests and footstools will also be beneficial to the desk worker.

Pheasant (1991) recommends the following suitable work heights.

- **Manipulative tasks** – those involving moderate force and precision to be 50–100 mm elbow height.
- **Delicate tasks** – from 50 mm to 100 mm above elbow height with wrists supported.
- **Lifting and handling** – should be from knuckle height to just above elbow height.

- **Two-handed pushing and pulling actions** – a little below elbow height.
- **Hand-operated controls** – between elbow and shoulder height.

Studies show typists and keyboard operators have benefited from adjustments to their work area. Nakaseko *et al.* (1985) noted that the redesign of keyboards allows more natural posture of the hands. The principles of this process incorporate a vertical split of the keyboard into two halves with an opening angle which avoids any ulna deviation, and a lateral slope to decrease the degree of pronation of the wrists. A split keyboard with a large forearm-wrist support was found to produce significantly more relaxation.

Shute and Starr (1984) tested the effects of 'advanced furniture' on a group of telephone operators. One experiment involved the use of an easily adjustable VDU table compared with a conventional non-adjustable one over an eight-week period. The second experiment involved the use of a conveniently adjusted chair, tested against a conventional one, over a five-week period. The results from both sides demonstrated less frequent discomfort, upper-back pain and ankle and foot discomfort. The advanced chair also led to reduction in shoulders and back discomfort.

Long-term studies to prevent RSI in the workplace, have proved that increase in productivity can occur. Westgaard and Aaras (1984, 1985) carried out research in a telephone exchange cable manufacturing plant in Norway, between 1975 and 1982.

Epidemiological health data was analysed and a comparison of the findings made with national averages suggested that long-term sick leave was associated mainly with musculoskeletal illness. Turnover between the period mentioned was as high as 30% per year. The researchers then looked at work duties and information relating to sick leave of more than three days. Respondents were also interviewed about the muscular pain intensity that they had experienced.

During 1975, alterations were introduced into the workplace. They involved the introduction of arm rests for sitting and manipulative tasks, altering lighting and providing new equipment. A couple of years later, the plant introduced flexible working hours. Westgaard and Aaras noted that electromyographic recordings of worker's trapezius muscles undertaken during work showed that alterations in the work environment reduced load by 50%. Sick-leave statistics and labour turnover was significantly reduced and productivity increased. Shoulder and neck pain disorders, once experienced, were also significantly reduced.

Modifications in lighting, work area dimensions, colour scheme and the provision of adjustable chairs, tables and foot rests, all contribute to the reduction in body fatigue and RSI according to Ong (1984).

A study conducted in Singapore among VDU workers saw the introduction of a work-rest regime, education relating to work habits and the promotion of provision of foot rests all related to the improvement of worker health. Prior to the experiment, VDU operators had a high prevalence of neck, shoulder and arm complaints.

THE USE OF TOOLS

Research has indicated that there is a need to:

- use the correct tools for the job being undertaken;
- ensure that the tools are suitable for the individual – the right size, good grip handle;
- avoid excessive manual force as it leads to unnecessary strain on the upper limbs;
- ensure tools are regularly serviced and in good condition;
- encourage the use where necessary of well-fitting, padded gloves with a good grip.

POSTURE

As recommended by ergonomists and physiotherapists, some simple straight forward rules may be followed to maintain good posture, thus reducing the risk of RSI:

- when sitting, ensure the back is straight;
- avoid reaching outside a comfortable range;
- when standing, keep as close to the work being done as possible;
- avoid sitting by the VDU for the entire day; take frequent breaks and supplement other clerical tasks for constant VDU work – this will allow the fingers and wrists to rest.

Managers should also ensure that training in manual handling should be undertaken by new employees when starting a new job. Updating of technique should be encouraged throughout the duration of employment. Managers should also be aware of current legislation and guidance relating to manual handling techniques and the prevention of injuries in the workplace.

LEGISLATION

There is currently no specific legislation (in the UK) that accommodates repetitive strain injuries. However, recent European Community Directives on health and safety in the workplace provide guidance to employers to reduce injuries.

1. Management of Health and Safety at Work Regulations 1992

The guidelines indicate that managers are to undertake:

- risk assessment;
- implementation of alterations relating to organization, planning, control and evaluation;
- the appointment of trained, competent and reliable individuals to assess, control and apply any alterations;
- the setting up of emergency procedures;
- provision of information to employees concerning health and safety matters.

2. Workplace Health, Safety and Welfare Regulations 1992

The regulations include ergonomic components such as:

- the suitability of work areas and seating;
- space and room dimensions;
- ventilation, temperature and lighting.

3. Provision of Use of Work Equipment Regulations 1992

For RSI prevention, special reference is made to the use of tools, transport, installation of machines and maintenance. All machinery must be safe and in good working order, suitable for the job and well maintained. Training is also indicated.

4. Personal Protective Equipment at Work Regulations 1992

Selection, provision, maintenance and the use of personal protective equipment such as protective clothing, hand, eye and foot protection are outlined in these directives.

5. Manual Handling Regulations 1992

Lifting operations (as outlined in Chapter 3) in the workplace are now ergonomically assessed by weight, shape and size of load. Tasks to be carried out, dependent on the individual's ability and capacity, are also highlighted. This identifies preventive measures against RSI and lower back injuries.

6. Health and Safety Display Screen Equipment Regulations 1992

As identified above, the use of VDUs can lead to upper limb disorders, eye strain, stress and tenosynovitis. It outlines the necessity for frequency of rest breaks and varying activities to prevent RSI.

SUMMARY

RSI is a broadly defined condition that has been in existence for many centuries. It is a disorder affecting upper limbs, hands, fingers, neck and back. RSI is treatable with rest, support and physiotherapy. It may be prevented if workers obey the rules relating to workspace, height adjustment of furniture, personal comfort and posture, as well as using the equipment with care.

Managers are bound by certain guidelines to ensure that the workplace is as safe and healthy as is reasonably practicable. However, specific legislation which could help to prevent RSI, does not yet exist.

SUGGESTED FURTHER READING

Cakir, A., Hart, D.J. and Stewart, T.F.M. (1980) *Visual Display Terminals*, John Wiley & Sons, Chichester.

Hunter, D. (1978) *The Diseases of Occupations*, Hodder & Stoughton, London.

REFERENCES

Armstrong, T.J., Foulke, J.A., Joseph, B.S. and Goldstein, S.A. (1982) Investigation of cumulative trauma disorders in a poultry processing plant, *American Industrial Hygiene Association Journal*, **43**, 103–16.

Buckle, P.W. and Baty, D. (1986) Ergonomic aspects of tenosynovitis and carpel tunnel syndrome in production line workers. In D.J. Oborne (ed.), *Contemporary Ergonomics*, Taylor & Francis, London, 237–41.

Cakir, A., Hart, D.J. and Stewart, T.F.M. (1980) *Visual Display Terminals*, John Wiley and Sons, Chichester.

Dainoff, M.J. (1982) Occupational stress factors in visual display terminal operation: a review of empirical research, *Behaviour and Information Technology*, **1**, 141–76.

Elias, R., Cail, F., Tisserand, M. and Christmann, M. (1980) Investigations in operators working with CRT display terminals; relationships between task, content and psychophysiological alterations. In E. Grandjean and E. Vigliani (eds), *Ergonomic Aspects of Visual Display Terminals*, Taylor & Francis, London, 211–17.

English, C.J., Maclaren, W.M., Court-Brow, C., Hughes, S.P.F., Porter, R.W., Wallace, W.A., Groves, R.J., Pethick, A.J. and Soutare, A. (1989) *Clinical Epidemiological Study of Relations Between Upper Limb Soft tissue Disorders and Repetitive Movements at Work*, Report TM/88/19, Institute of Occupational Medicine, Edinburgh.

Grandjean, E. (1984) Postures and the design of VDT work stations, *Behaviour and Information Technology*, **3**, 301–11.

Grieco, A. (1986) Sitting posture: an old problem and a new one. *Ergonomics*, **29**, 345–62.

Hult (1954) Cervival, dorsal and lumbar spinal syndromes, *Acta Orthopaedica Scandinavia*, Supplement 17.
Hunting, W., Grandjean, E. and Pinderman, M. (1983) VDT Workstation Design: Preferred Settings and Their Effects, *Human Factors*, **25**, 193–6.
Maeda, K., Havada, N. and Takamasta, M. (1980) Factor analysis of complaints of occupational cervicobrachial disorder in assembly lines of a cigarette factory, *Kurume Medical Journal*, **27**, 253–61.
Maeda, K., Horiguch, S. and Hosokawa, M. (1982) History of the studies on occupational cervicobrachial disorder in Japan and remaining problems, *Journals of Human Ergology*, **11**, 17–29.
Nakaseko, M., Grandjean, E., Hunting, W. and Gierer, R. (1985) Studies on ergonomically designed alphanumeric keyboards, *Human Factors*, **27**, 175–87.
Nishiyama, K., Nakaseko, M. and Uehata, T. (1984) Health aspects of VDT operators in the newspaper industry. In E. Grandjean (ed.), *Ergonomics and Health in Modern Offices*, Taylor and Francis, London, 113–18.
Novek, J., Yassi, A. and Spiegal, J. (1990) Mechanization, the labour process and injury risks in the Canadian meat packaging industry, *International Journal of Health Services*, **20**(2), Baywood, USA, 281–96.
Ong, C.N. (1984) VDT Workplace design and physical fatigue: a case study in Singapore. In E. Grandjean (ed.), *Ergonomics and Health in Modern Offices*, Taylor & Francis, London, 484–94.
Pheasant, S. (1991) *Ergonomics, Work and Health*, Macmillan, London.
Ramazzini, B. (1713) *De Morbis Artificum*. Translated by W.C. Wright (1940) *Diseases of Workers*, University of Chicago, Chicago.
Roto, P. and Kivi, P. (1984) Prevalence of Epicondylitis and Tenosynovitis among Meat Cutters, *Scandinavian Journal of Work, Environment and Health*, **10**, 203–5.
Shute, S.J. and Starr, S.S. (1984) Effects of adjustable furniture on VDT users, *Human Factors*, **26**, 157–70.
Smith, M.J., Stammerjohn, L.W., Cohen B.G.F. and Lalich, N.R. (1980) Job stress in video display operations. In E. Grandjean and E. Vigliami (eds), *Ergonomic Aspects of Visual Display Terminals*, Taylor & Francis, London, 201–10.
Viikari-Juntura, E. (1984) Tenosynovitis, peritenditis, and tennis elbow syndrome, *Scandinavian Journal of Work, Environment and Health*, **10**, 443–9.
Wallace, M. and Buckle, P. (1991) Ergonomic aspects of neck and upper limb disorders, *International Reviews of Ergonomics*, **1**, Taylor & Francis, London, 173–200.
Westgaard, R.H. and Aaras, A. (1984) Postural muscle strain as a causal factor in the development of musculoskeltal illnesses, *Applied Ergonomics*, **15**, 162–74.
Westgaard, R.H. and Aaras, A. (1985) The effect of improved workplace design on the development of work related musculoskeletal illnesses, *Applied Ergonomics*, **16**, 91–7.

USEFUL ADDRESSES

The Ergonomics Society, Devonshire House, Devonshire Square, Loughborough, Leicestershire LE11 3DW.
Napier University – Applied Ergonomics Unit, Department of Biological Science, Colinton Road, Edinburgh EH10 5DT.

Robens Institute of Ergonomics, Research Unit, University of Surrey, Guildford, Surrey GU2 5XH.

NOTES

1. Oedema refers to an effusion of fluid in the tissues.
2. Fibrosis is literally the formation of fibrous tissue as a result of scarring or inflammation.
3. Crepitus is the grating sound caused by friction of bones.

Managing health at work

9

Introduction • Health risk appraisal • Risk assessment and risk management • Considering ways of combating difficulties • Monitoring developments and improvements • Awareness of current legislation • Conclusion • Suggested further reading • References

INTRODUCTION

Managing health at work is a task that will need the support of senior management, not only in terms of resources, but concerted action to enable organizational and employee well-being. Improved productivity, reduced sickness and absenteeism, provision of effective and efficient services, are just some of the benefits of a healthy organization. However, the manager cannot be the sole driving force to improve employee health. Health is everyone's business: the groundsmen, the secretary, the catering team, the personnel officer. Utilizing the staff resource in the most effective way possible will enable the organization to achieve its goals.

People need to be guided, and at the same time to have some sense that they are achieving their own aims whether it be personal fulfilment or career development. They will need to feel that they are participating in the organization's initiatives, have communicated their ideas, perhaps even contributed in some aspects of planning, implementation and evaluation of the work they have been involved in. Research has proven that worker participation contributes to the relief of unwanted stress and anxiety about work and employee's specific roles (Deutsch, 1981; Gardell, 1987; Levi, 1981; Gardell and Gustavsen, 1980).

Managing health at work will need to take these issues on board if success and achievement that benefits all is to result. In order, to begin this process of worker involvement and workplace health improvement, the manager should:

1. identify the problems and issues that contribute to worker ill health;
2. consider ways of combating the difficulties and coping with the main issues concerned;
3. formulate a strategy for health improvement that encompasses organizational and employee needs;
4. monitor developments and improvements;
5. keep abreast of current legislation.

IDENTIFYING PROBLEMS AND ISSUES

- Do the problems relate to productivity?
- Are there issues relating to sickness and absenteeism? How are they being dealt with at present?
- Have new materials/equipment been introduced into the organization that are unfamiliar to the workforce or that are potential hazards?
- Are there issues relating to work design leading to boredom or repetition or role ambiguity?
- Are there problems with support systems for colleagues in the organization?
- Is smoking within the organization leading to conflict amongst smokers and non-smokers?
- What contributions has the organization already made towards improving employee health that have been relatively successful? How were they achieved? What lessons have been learned ? How can these ideas be carried forward?

The human resources department within the organization have limited time and resources available to deal with all the issues that relate to employee health. With the exception of large organizations, few companies have occupational health facilities. Those fortunate enough to have them focus on treatment and care of the injured as opposed to prevention (although prevention in occupational healthcare is a developing area). Currently, any initiative to improve worker health tends to be left to the enthusiasm of the occupational health nurse and any persuasive arguments s/he can put to the senior management team. The occupational health team needs greater support.

The human resources and occupational health teams may be regarded by the workforce as agents of management. The effect of any initiative to improve worker health may thus be seen as just another gimmick or a means of controlling or restricting worker activity.

Workplace health initiatives can be achieved with co-operation between managers and the workforce working in unison. Any initiative needs co-ordination. This can be done objectively by employing the services of a health promotion officer with a workplace health brief, for a limited period. However, the role is envisaged as a developmental one and it may be beneficial in the long term to fully employ a workplace health co-ordinator who will become familiar with the specific needs of the organization and its workforce. Alternatively, there may be a decision to utilize existing resources within the company. Whatever decision is taken, some understanding of the role and functions of the co-ordinator will be necessary.

DEVELOPING ROLE OF THE WORKPLACE HEALTH CO-ORDINATOR

The role of the workplace health co-ordinator is envisaged as a pivotal one for liaison between management and the workforce. The co-ordinator should report to a member of the senior management team who will have issues relating to workplace health as part of his/her portfolio of responsibilities.

The co-ordinator's role is likely to include:

- health risk appraisal: organizational and individual;
- research and development;
- policy development;
- advice and consultancy;
- evaluation;
- awareness raising;
- developing/delivering training programmes;
- networking with local and national support groups;
- marketing.

HEALTH RISK APPRAISAL

As noted in the chapters relating to cancer and repetitive strain injury (RSI), there are specific work practices that can lead to exposure to hazardous substances and preventable injuries.

Individual health risk appraisals were introduced during the 1970s in the USA. They were seen as an effective approach to screen for and provide individualized health education concerning particular health risks. They also gave the opportunity for employees to discuss health issues generally and to seek advice.

The process, still employed today, utilizes a computer package to obtain epidemiological and statistical data to provide individuals with projections of their personal mortality risk. Health-risk reduction strategies are discussed and employed as a result.

Identifying problems and issues at organization level will involve risk assessment.

RISK ASSESSMENT AND RISK MANAGEMENT

Risk management as defined by Dickson (1995) is a mechanism for managing exposure to risk that enables individuals to recognize the events that may result in unfortunate and damaging circumstances in the future, their severity and how they can be controlled. Briefly, it involves the identification, analysis and economic control of the assets or earning capacity of an organization. Risk management has its origins in the manufacturing and process industries where the process of manufacture was regarded as the key to workplace ill health, and general environmental problems. The principles of risk management and assessment are equally applicable to the service sector.

Dickson emphasizes that risk management should be seen as a positive way of assisting operational managers to achieve their objectives. It is through identifying, measuring and controlling risk that the assets, earning capabilities and the objectives of the organization and its business will be secured.

Risk assessment strategies were devised to locate potential and causative agents of ill health and injury. The assessment employs two methods: thematic and strategic assessment.

Thematic assessment involves an examination of certain work environment factors, e.g noise and physiological loads in the workplace. The problems and potential consequences are analysed and the manufacturing process is scrutinized to identify the cause. This type of assessment is used in detecting the causative agents, of for example, repetitive strain injuries (see Figure 9.1).

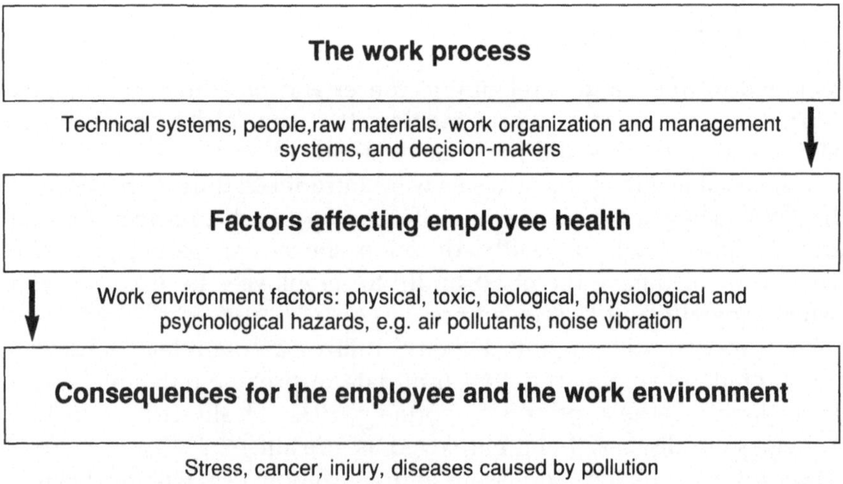

Figure 9.1 Process of thematic assessment

When undertaking thematic assessment of **the work process** the following considerations will be necessary.

- What is involved?
- Are the technical systems fluent? Checked regularly? Safe?
- Are the workforce sufficiently trained to achieve their maximum potential with the systems of operation?
- Are there adequate storage facilities for products?
- Are the workforce aware of any potential hazards of handling raw materials and the precautions to take?
- Are work and management systems sufficiently flexible for smooth operations and reduction of tension among workers?

Physical, toxic, biological, physiological and psychological hazards must be considered when noting **factors affecting employee health**.

- How frequently are noise levels and toxic emissions measured? Altered? Safe?
- Do employees have sufficient protective equipment and follow guidance?
- Do all employees know how to deal with an emergency?

What are the **likely consequences for employee health and the work environment**?

As mentioned in previous chapters, employees may be exposed to working conditions that could lead to stress, chest diseases, cancers and injury. Therefore, thematic assessment will be attempting to identify:

- Which policies and strategies do managers have for dealing with the likelihood of an employee contracting a particular disease or sustaining injury as a direct result of their work?
- How can the risk of litigation be minimized?

Strategic assessment is used to identify weaknesses in the management structure within the organization of work tasks and possibly in the co-ordination of health and safety, as well as overall policy in the organization. Such weaknesses may include:

- poorly defined goals;
- lack of control in purchasing;
- inconsistent means of identifying and examining problems;
- poor maintenance;
- defective communication systems (see Figure 9.2).

Within the context of workplace health, thematic and strategic assessment will be useful in identifying the potential causes of strain, excessive stress and the prevention of exposure to chemicals, as well as

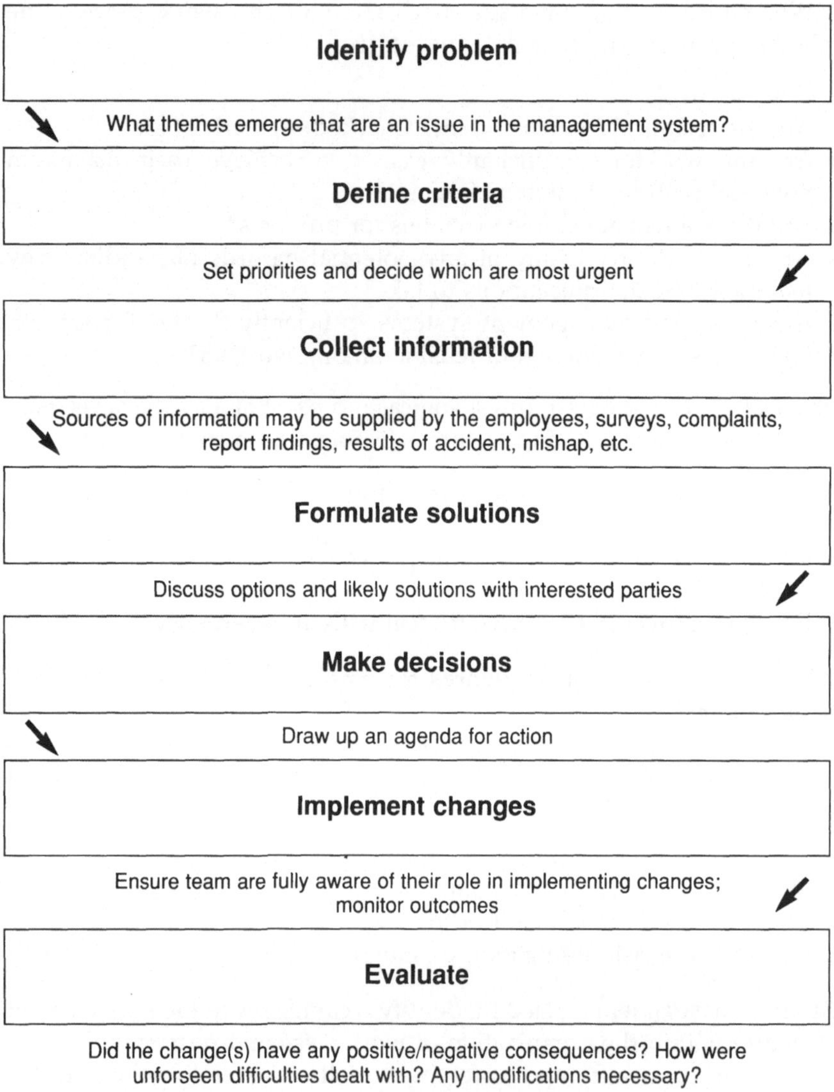

Figure 9.2 Process of strategic assessment

the problems relating to management and communication systems throughout the organization.

Part of the workpace health co-ordinator's role will involve some aspect of **marketing.** The co-ordinator will need to provide up-to-date, succinct and useful information that is suitable to the workforce. Understanding of the basic fundamentals of marketing and how to make an effective 'sale' will be useful in initiatives such as open days and health fairs. Care and consideration of the selection of suitable materials with appropriate health messages is a vital part of this role as the co-ordinator will be supplying marketing materials to the caterer, fitness instructor, occupational health department. The company may also wish to develop a corporate style, thus knowledge of the organization's needs and the image it wants to portray to the public are areas the co-ordinator should discuss with the senior management team. Gaining access to the press and television networks will also be a part of the role of the workplace health co-ordinator.

The workplace health co-ordinator will also be involved in the **development and delivery of training programmes**. Being aware of developments in the workplace health field is important in taking any initiative forward. For example, to set up a healthy-eating initiative, caterer and catering managers will need to be informed of the latest innovations in the food industry, relevant policy and menu planning for health, as well as developments relating to employee incentives.

The initiative is likely to contribute towards the altering of the organization's culture, therefore getting the foundations in place first is the wisest approach. Naturally, the co-ordinator is not expected to be an expert in every aspect of health, but can support others in their initiatives and make use of experts in the field.

Managers will need to know if workplace health initiatives are working well or if modifications are necessary. **Evaluation** of outcomes to assess the benefits of the venture in terms of health and cost effectiveness will also be seen as part of the workplace health co-ordinator's role.

Other workplaces in the local area or subsidiary companies are likely to call upon the services of the workplace health co-ordinator if s/he proves to be a success within the company. Acting as **consultant** to other companies may prove valuable in forming healthy alliances.

CONSIDERING WAYS OF COMBATING DIFFICULTIES

To consider ways of combating problems and issues identified within the organization through risk assessment and health-risk appraisal, the workplace health co-ordinator will need to form a working party.

The working party is likely to consist of:

- the workplace health co-ordinator;
- union representative;
- personnel officer;
- occupational health nurse;
- other interested parties, e.g quality officer, catering manager, physiotherapist;

- external expertise, e.g nutritionist, employee assistant programme adviser, counselling services representative.

The working party will discuss the issues, formulate an action plan for tackling them, and implement any changes necessary. The issues will vary, for example formulating a new no-smoking policy or coping with an employee with known alcohol abuse problems will require differing knowledge and expertise, such as clarity on the legal position; whereas, a healthy-eating initiative may require the presence of a Gardener Merchant catering expert with long-standing experience in the field.

For further guidance see Figure 9.3.

Other ways of considering and combating workplace health issues involve the use of local support networks. These will include statutory and voluntary organizations who may be able to assist the company

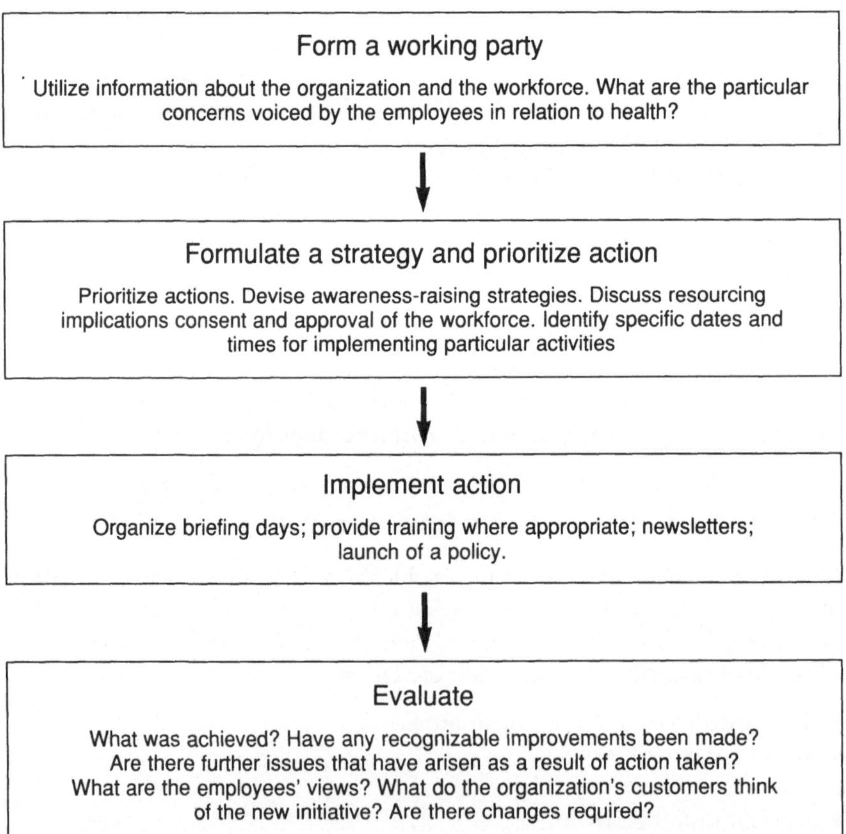

Figure 9.3 Workplace Health Action Plan

from time to time because of their particular knowledge or expertise. These will include:

- leisure centres and health spas;
- health food suppliers;
- aromatherapist, physiotherapist, Alexander technique instructor;
- Health and Safety Executive;
- voluntary agencies such as MIND, BACUP, Cruise, RELATE, ASH, Alcohol Concern, National Backpain Association;
- Health At Work Business Unit;
- Institute of Personnel Development;
- companies that have launched a recent initiative.

Learning from others is a valuable way of identifying the types of information needed and techniques for carrying initiatives forward. Indeed, Scandinavia leads the world in research on occupational stress and has used research to make a major contribution to social policy and workplace health reforms. Such contributions as the 1977 Norwegian Work Environment Act and the 1978 Swedish Environment Act are still applicable today. Under Section 12 of the 1977 Act:

> (1) General Requirements
> Technology, organization of the work, working hours and wage systems shall be set up so that the employees are not exposed to undesirable physical or mental strain and so that their possibilities of displaying caution and observing safety measures are not impaired.
>
> (2) Arrangement of Work
> The individual employee's opportunity for self-determination and professional responsibility shall be taken into consideration when planning the work.
>
> Efforts shall be made to avoid undiversified, repetitive work and work that is governed by machine or conveyer belt in such a manner that the employees themselves are prevented from varying the speed of the work. Otherwise efforts shall be made to arrange the work so as to provide possibilities for variation and for contact with others, for connection between individual job assignments and for employees to keep themselves informed about production requirements and results.
>
> (3) Control and Planning Systems
> The employees and their elected union representatives shall be kept informed about the systems employed for planning and effecting the work, and about planned changes in such systems.

> They shall be given the training necessary to enable them to learn these systems, and they shall take part in planning them.
>
> (Deutsch, 1981)

A crucial part of these legislative agreements has been worker participation. Worker participation leads to improved job health and safety, and stress reduction. Management strategies that incorporate the participative mode will assist in general improvement. The workplace health co-ordinator's function will be vital in this capacity in ascertaining the opinions of the workforce as well as identifying their needs and including them in the planning, implementation and evaluation of initiatives.

MONITORING DEVELOPMENTS AND IMPROVEMENTS

Senior managers within the organization will also need to be kept informed of developments relating to workplace health improvements. It is therefore advisable that a member of the senior management team reports to her colleagues at board level.

The workplace health co-ordinator, in conjunction with her colleagues will need to continue the process of internal collaboration, monitoring sickness/absence, finding out how employees are coping as a result of alteration in their work design, discover whether the no-smoking policy is working and that domestic staff are not given an additional workload cleaning up cigarette ends not usually found in otherwise clean parts of the building, keeping abreast of developments in infection control, monitoring accidents. Making use of internal intelligence will further identify any changes necessary in the management and information systems.

AWARENESS OF CURRENT LEGISLATION

In the UK, employees are protected under a series of legislative agreements which outline their rights and obligations as employees in relation to safety and health. Managers are responsible for ensuring that compliance with legislation is upheld by their company and that regulations are enforced as far as possible.

The present list includes:

- Health and Safety at Work Act 1974;
- COSHH Regulations 1988 and Amendments 1994;
- The Management of Health and Safety at Work Regulations 1992;
- Workplace Health, Safety and Welfare Regulations 1992;
- Provision of Use of Work Equipment Regulations 1992;
- Manual Handling Regulations 1992;

- Personal Protective Equipment Regulations 1992;
- Health and Safety Display Screen Equipment Regulations 1992.

CONCLUSION

Workplace health improvement involves utilizing available resources, making use of the skills of the workforce, establishing support networks and keeping abreast of current developments in legislation. For the continued health of the whole organization, co-ordination of workplace health initiatives is necessary, as well as rigorous monitoring of potential hazards, sickness and absenteeism, and measuring these elements against productivity. Continuous research is required into problems affecting worker health within individual organizations. This can be assisted by proactive managers who will support and publicize issues, and successes as they arise. It is only through this continuous process of monitoring, research and development that advancement can ever be made in improving workplace health in terms that are beneficial to all.

SUGGESTED FURTHER READING

Gemson, D.H. and Sloan, R.P. (1995) Efficacy of computerised health risk appraisal as part of a periodic health examination at the worksite, *American Journal of Public Health Promotion*, **9**(6), 462–6.

Paoli, P. and Litske, H. (1994) *Monitoring the Work Environment*, Report of the Second European Conference. European Foundation for the Improvement of Living and Working Conditions.

Hopkin, P. (1995) A new approach to risk management, *Health and Safety at Work*, October, 8–10.

REFERENCES

Deutsch, S. (1981) Work environment reform and industrial democracy, *Work and Occupations*, **8**(2), 180–94.

Dickson, G. (1995) Principles of risk management, *Quality in Health Care*, **4**, 75–9.

Gardell, B. (1987) *Work Organization and Human Nature: a Review of Research on Man's Need to Control Technology*, Swedish Work Environment Fund, Stockholm.

Gardell, B. and Gustavsen, B. (1980) Work environment research and social change: current developments in Scandinavia, *Journal Occupational Behaviour*, **1**(1), 3–7.

Levi, L. (1981) *Preventing Work Stress*, Addison-Wesley, Reading, Massachusetts.

Index